THE INDEPENDENT HOME

p.16 whole life cost analysis
p.28+ retrofitting priorities
p.54-55 graph of PV prices
solar politics (CIs, etc.)
p.77 good battery box design
p.87 neat batteries

THE REAL GOODS
INDEPENDENT LIVING BOOKS

Paul Gipe, *Wind Power for Home & Business: Renewable Energy for the 1990s and Beyond*

Michael Potts, *The Independent Home: Living Well with Power from the Sun, Wind, and Water*

Real Goods Trading Company in Ukiah, California, was founded in 1978 to make available new tools to help people live self-sufficiently and sustainably. Through seasonal catalogs, a quarterly newspaper (*The Real Goods News*), an annual *Alternative Energy Sourcebook*, as well as a book catalog and retail outlets, Real Goods provides a broad range of renewable-energy and resource-efficient products for independent living.

"Knowledge is our most important product" is the Real Goods motto. To further its mission, Real Goods has joined with Chelsea Green Publishing Company to co-create and co-publish the Real Goods Independent Living Book series. The titles in this series are written by pioneering individuals who have firsthand experience in using innovative technology to live lightly on the planet. Chelsea Green books are both practical and inspirational, and they enlarge our view of what is possible as we enter the next millennium. The authors of these Independent Living Books light the path ahead of all of us.

Ian Baldwin, Jr.
President, Chelsea Green

John Schaeffer
President, Real Goods

THE INDEPENDENT HOME

Living Well with Power from the Sun, Wind, and Water

MICHAEL POTTS

A Real Goods Independent Living Book

CHELSEA GREEN PUBLISHING COMPANY
POST MILLS, VERMONT

Design by Ann Aspell
Ornament by Rene Schall

Printed in the United States of America.
1 2 3 4 5 6 7 8 9 10

Potts, Michael, 1944–
The Independent Home : living well with power from the sun, wind, and water / by Michael Potts.
p. cm. — (A Real goods independent living book)
Includes bibliographical references and index.
ISBN 0–930031–65–2 (pbk.)
1. Dwellings—Energy Conservation. 2. Renewable energy sources.
I. Title. II. Series.
TJ 163.5.D86P68 1993
696—dc20 93–11665

Chelsea Green Publishing Company
P.O. Box 130, Route 113
Post Mills, Vermont
05058-0130

Due to the variability of local conditions, materials, skills, site, and so forth, Chelsea Green Publishing Company and the author assume no responsibility for personal injury, property damage, or loss from actions inspired by information in this book. Always consult the manufacturer, applicable building codes, and the National Electric Code™ before installing or operating home energy systems. For systems that will be connected to the utility grid, always check with your local utility first. When in doubt, ask for advice; recommendations in this book are no substitute for the directives of equipment manufacturers or federal, state, and local regulatory agencies.

To Sienna and Damiana, my daughters,
and Rochelle,
who remind me of the purpose

Contents

FIGURES

Foreword

In 1971 I graduated from Berkeley and left campus with a used tent and a North Face sleeping bag in tow for the wilds of Mendocino County in northern California. I suspected that life would never be the same again. My 8-foot-square, $50 tent—with a kerosene lamp for a power system—became my first independent home.

Seven years and several hundred 80-mile commutes to work later, I discovered some old lightbulbs marked "12-volt" gathering dust on the shelf in a local hardware store in Ukiah. My curiosity piqued, I brought the bulbs home to my latest independent home, which by then was a 12-by-18, single-walled redwood cabin. To my joy and amazement I discovered that the 12-volt lightbulbs could be operated with the juice from my car battery—and in one fell swoop the twentieth century arrived at my humble dwelling. Now I could feel good about my polluting commute because the car had been transformed from a mere gas-guzzler into a battery charger for my house. Before long I was entering the consumer age with every ugly appliance the RV industry could churn out in 12-volt, from blenders to 12-inch TVs. I became the techno-radical in a community of neo-Luddites. Little did I realize that I was tinkering with the rudiments of what was to become "the independent home movement."

In 1978 I opened a store in Willits, California, called Real Goods Trading Company, to serve all the other urban refugees who like me were experimenting with independent homes and power systems. We kept thinking that there just had to be a better way to charge batteries than driving around in circles or firing up smelly diesel generators. Then one day, a guy walked in the store and offered to sell me a large quantity of the first photovoltaic modules (PVs, or solar-electric panels) I'd ever seen, for about $100 per watt. They were outrageously expensive but irresistible, promising an absolutely new way to electrify the wilderness, and judging from our market at the time, price was not a particular obstacle. So I bought what were to become the first photovoltaic modules sold retail in America.

I remember a visit by reporters from the *Manchester Guardian* who wanted to know if it was just the "pot growers" who were using solar for their gardens. The dominant thinking at the time was that solar was the pipe dream of environmentalists and hippies, and that this complicated and costly new technology would never amount to much more than an inventor's plaything. Northern California quickly became the petri dish for the study of solar-electric energy, as one oil company executive after another lear-jetted into town to observe this bizarre phenomenon: PVs were selling as fast as we could supply them to the latest influx of ex-urbanite settlers.

In the fifteen years since, nothing short of a renewable-energy revolution has occurred, spreading from the hills of Mendocino County all the way to the center of power in Washington, D.C., where Real Goods is presently involved in a feasibility study to design solar for the White House. Indeed, solar-electric power has come of age. It has become for many homeowners just another invisible source of electricity, but one that doesn't pollute, make noise, or rack up monthly bills, and one that works year-in and year-out. There are now more than 100,000 homes in the U.S. that derive their electricity from photovoltaics.

But what about the people in those homes? PV is only a technology. The real story of this revolution is the story of the individuals and families who have pioneered the technology, who have been brave enough to sever their connection with the utility grid and declare the independence of their homes and livelihoods. The spark that drives someone to just say no to the utility company is mysterious and irrepressible—it begs to be better understood. And the power of unplugging cannot be underestimated. What is it that compels a person to break away? Can anyone do it?

Michael Potts has done a truly commendable job of bringing clarity to the story of the renewable energy revolution: he has integrated for us his insights into the technical magic of solar with his sense of the humanity of those who have lived with the new technologies. I first met Michael thirteen years ago while he was browsing among the PV panels at our Ukiah store. I've watched him take his own beautiful, eclectic, coastal independent home off the grid little by little over the years, like a work of art slowly assembling. Michael wrote the computer programs that were the foundation of the Real Goods mail order entry system, and he has been singularly enraptured by the exponential growth of renewable energy use in this country. A frequent book reviewer and contributor to the *Real Goods News* as "the Curmudgeon," Michael has also written our *Remote Home Kit Owners' Manual.*

To write *The Independent Home* he criss-crossed the U.S. and broke bread with the pioneers who have left the grid behind. He gives you their stories, which are real, not the stuff of fantasy; you'll see that the exhilarations and joys of living in an independent home are not without their attendant tribulations. Most important, Michael reveals and elucidates the shared spirit he encountered in these new pioneers, a spirit which is now part of the construction of his book and which gives *The Independent Home* its striking vitality.

The new solar technologies and the "independent home spirit" are contagious. Witness the circumstances of this book's writing and production: from the writer and the editor to the designer and the publishers (including myself), all of us have been taken by the spirit—and have converted or are making plans to convert our homes to independent power systems. *The Independent Home: Living Well with Power from the Sun, Wind, and Water* will show you how to join us.

John Schaeffer
President
Real Goods Trading Company

Introduction

JUST ABOVE THE ROAD FROM CASPAR TO UKIAH THERE IS A SPRING OF LEGENDary sweetness and stability. The Yuki Indians stopped here on their annual trek from the inland valley to the coast, and stopped again on the way back. Local folks come and park their pickups beside the spring and fill jugs from the crystalline flow. Not many years ago, they brought their garbage along, and heaved it all—batteries, paint thinner, tires, disposable diapers—over the downhill side, where the spring's overflow trickles down to the river and the sea. Yet in the last decade, an amazing leap of consciousness has taken place. Not only has the dumping stopped but the community has spontaneously come together to clean up the midden. It will be decades before animals can drink safely from the spring's overflow, but somehow we reached a collective understanding: We must not dump where we drink!

For me, this small event is an emblem of a greater rebirth: the return to the home place.

In the sweep of history, the twentieth-century American house will probably be regarded as a temporary aberration, an embarrassment to enlightened builders and planners. It will be called "the out of place house" or "the utterly dependent house." During this century, the American housing industry developed a house perfectly disconnected from its environment, the needs of its inhabitants, and the true notion of homeplace. These cookie-cutter houses filled with resource-gobbling conveniences are dropped on the land for the convenience of subdividers and builders, rather than placed to take advantage of sun, wind, and storm. Such marvels as pure water, electricity, and fuel for heating and cooking come through pipes and wires; most of us approach the sources of supply most closely when we pay our utility bills. Our waste products are dealt with as conveniently: down the drain or chute, out the window and up the chimney. Many of us went to school at a time when rivers, oceans, all the resources we depended on, were still thought to be boundless, and our mother earth seemed infinitely giving and forgiving. Humankind has been, for a brief moment in time, absolute ruler and shaper of the planetary domain, as we came to believe that we could do our will with the species and elements at hand. The explosive development of consumerism, the replacement of nature by technology, and the quest for wealth through disposability and obsolescence have transfixed us and deadened us to the rhythms and powers of our partners on the planet.

Still the sun shines on! Many of us noticed this disconnectedness from nature, and yearned for a comfortable, self-sustaining living space; we have given our whole lives to reinventing what should be the most natural thing in the world: a home. We have begun a simple, healing movement toward a home place where we can enjoy the best of our modern lifestyle, knowing that our comfort is not taken at the price of another's misery; where we can relearn the natural cycles that rule all life on our planet; and where we can widen our vision to include the whole wondrous globe we share.

Ecology, you remember, is the study of the ecosystem, from the Greek οικοζ, meaning *home*. This book considers the homeliest application of that science, the system of the home place. If we look deeply enough, we discover that in nature all systems are closed, which means that what we consume today will not be available tomorrow (depletion) and what we discard today will come back to plague us tomorrow (persistence). By applying this principle to the home, we have founded a new frontier.

Living with a house: Dennis Weaver's story

As eloquent as home places are, it is those who have built them and inspire them with life who can best explain the choices they have made, the right and wrong paths they have followed, their compromises and triumphs and discoveries. These are pioneering voices, even though their frontier was commonplace not more than a century ago, and remains so on every other continent. Part of the frontier, and the most interesting aspect of this book, is the struggle to bring global consciousness into focus on the most local of all places, our homes.

Dennis Weaver divides his time between an earthship—a sun harvesting home built with earthen walls, discarded tires and cans, and local materials into the south slope of the foothills near Ridgway in southwestern Colorado—and a conventional house in California. Let him tell you a story about the houses he lives in.

Our main house is our earthship in Colorado. We've been there for three years, and I'd say we spend half our time there. The thing that attracted me about the earthship is that it is environmentally friendly, it is self-sufficient, it is energy-efficient, and it uses hardly anything from the conventional systems that make a conventional house functional. We aren't tied to the grid. We aren't plugged in, you might say, like in a hospital, when you are plugged into life supports. The conventional houses that we live in now are on life support.

There's a very natural feeling in an earthship. The warmth comes from stored energy, stored temperature in the mass of the house, and it

Dennis Weaver's earthship home in Colorado is made with discarded tires filled with earth, empty cans, and local timber (photo courtesy of Alice Billings).

just naturally releases at night to keep the air temperature warm or cool. There's no heating ducts or air conditioning, so you don't have artificial temperature control, and you don't have a lot of sound going on which is man-made and unnatural. We have room to grow a lot of plants in the house, and they grow so beautifully because it is a natural environment for them. It feels good, the air is real good. Of course if it warms up in the house, we've got the capacity to control that by opening skylights, windows, and you get a nice moving air feeling. And it's very natural moving air; it's not like it's pushed out with a fan.

It's arid in Ridgway, and there's a lot of sage brush over here, but the soil is great. We have a meadow down below us where the llamas graze. There's a lot of rainfall and a lot of snow. We have a little forest behind our house and to the side of it, with all kinds of trees, pine trees and piñon trees, a couple of cedars and ponderosas. We have a garden which is very, very productive.

You have to live with the house. You know, your house is your partner, it's your companion, it's your shelter, and you have to live with it in that sense. You can't just go over to the wall and set a thermostat, then go off and leave it and get all this artificial stuff going. You have to know that when it gets warm, you open windows and let the heat out, and you have to know that when night comes and it's wintertime, you need to close the windows as the sun is going down so that you hold the warmth in the house. You just have to live with it. I don't feel that's a drawback. I think it's really nice.

Moving back and forth between an on-the-grid and an off-the-grid house, you have to make an adjustment. It's not at all hard. I get annoyed with the California house, because sometimes the thermostat gets screwed up, and about three o'clock in the morning the heat goes on . . .

When we're in southern California, and when the electricity goes out, our plug is pulled, and we're helpless, totally helpless: we're at the mercy of the central power system. That's not the case out in Colorado, where the power can go out in the neighborhood around us and we don't even know it. I'd certainly like to install solar power in the California house. That would be terrific, because then we wouldn't be a victim to the system, and it also would be a clean source of power, an inexhaustible source of power, and a cheap source of power. So I'd really like to do that.

The Ridgway experience has just pointed out the fact that we are held captive by the systems, and that we use a lot of energy in our house that comes from dirty sources, coal-fired powered plants or nuclear power, and we are very aware of that. We're also very aware that our power is coming from a limited source. It's something that is not going to last. I may not outlive the fossil fuels, but certainly my grandchildren will outlive them. And I think we should be thinking about our grandchildren. Some scientists are saying we only have something like a ten-year supply of petroleum left. I don't know if that's true or not, but let's say it's fifty or sixty: it's going to be gone. Whatever we're going to be forced to do at that time, it just seems sensible to me to start doing it now. Don't wait until the crisis is on us. Let's manage by objective rather than crisis.

There is no question: my major concern is for the place where we live, and the environment where we all have to have to live. I mean, every living thing needs a proper environment in which to exist, and if it doesn't have it, it dies. Human beings are no different. And the sad part, the absolutely desperate part of what we're doing is, we're destroying our environment, we're destroying our life support. And that is something that I'm extremely concerned about. I talk about it all the time. I go out on the lecture circuit and talk about it, and I try to incorporate solutions in my own lifestyle.

I also have an electric car. They're at a primitive stage right now, but that industry needs to be encouraged. I use it for short trips. You know, we should support that industry because it's a thing of the future, or one of the possibilities anyway, and electric vehicles reduce the amount of energy we use. It's the energy that we consume that really creates the environmental problems we're dealing with, whether it's acid rain or depletion of the ozone layer. We're using a dirty source, we're using an exhaustible source, we're using an expensive source. When you consider that we are held hostage, and have been in the past, by some crank across the ocean, because our energy source, petroleum, is threatened—we've got to get away from that. We've got to be free of that energy source for our own national security, and for our own healthy economy; we've got to do it. We have no choice there.

That's not going to be a tragic thing. I claim that there's a huge environmental industry waiting to be tapped, which will create a strong economic base for us, which will produce a tremendous number of jobs. You know, as human beings, we go through realignment all the time; we have to adjust. As we have evolved, we have continually adjusted. Just look what's happened through the industrial revolution: it's been nothing but one adjustment after another. I mean, the idea that adjustment is something we have to be fearful of is really silly. Of course industries will change and people will have to retrain, but when the automobile was invented, the farriers had the same reaction. What happened to the horse-shoers? They went into repairing cars. Listen, if we do not create and back an environmental industry in this country, we're going to be swept under the rug, just like we were with automobiles at one time, and computers. All that technology that we created is in the hands of foreign countries, and we're buying from them instead of producing. If we don't get in step with what is needed for the future, and go into an industry supported by another kind of energy source, we're going to be back there in the back of the pack someplace. The industry will be transferred overseas, and we will become, more and more, as we have in the past ten years, a service-oriented country.

I can't give any special advice; there's so much good advice out there, so many good books. Get a simple book, and watch your energy use. And be concerned. Be aware that we're going to have an immense problem if we don't address it now. I mean, it's already an immense problem, but if we continue to stick our heads in the sand and ignore it, live in denial, and say we don't have a problem, it's only going to get worse.

A Short Note on Methodology

This book is composed of stories like Dennis Weaver's; facts and explanations about earth-friendly houses and the systems that power them; and opinions distilled from my own experience as well as hundreds of conversations, one hundred and fifty questionnaires, and one hundred formal interviews with energy-conscious informants. Any group of Americans, and certainly a group as critically aware as this one, maintains a broad diversity of opinions, and on any particular point my own conclusions will differ from that of many of my informants. I have tried to stay within the spectrum of broadly conceived consensus, but my enthusiasm for my subject has led me, at times, into possibly outrageous opinionatedness. In this book few of the ideas, but all of the errors, are mine alone.

To gather information for the book I travelled twenty thousand miles, because I had to satisfy myself that this movement is happening everywhere. I spent hundreds more hours travelling electronically by telephone and computer network. Wherever I looked, I found extraordinary people who were doing something positive about their relationship to the earth and to energy. They were generous with their time and patient in their explanations. Transcribing what I was told, I "wrote myself out" and tried to record the voices and ideas of the storytellers faithfully; when a story jumps from topic to topic, it is probably because I diverted the teller with a question. Each storyteller tries to convey what she or he believes is most important for you to know about independent homesteading. As a result, it may seem that remarks made early on in the book lack background that will eventually be found in a subsequent chapter. For example, in Dennis Weaver's story, what, precisely, is an earthship? You will find out more about earthships in chapters 6, 8, and 14.

Pioneers may remember (and minimize) their hardships, but often they forget entirely the breadth of unknown territory they were required to explore to finally get home. While the stories will guide us, I will review concepts you may have forgotten and bring you up to date on the technology and the marketplace of energy independence. When you encounter a term you do not understand, the chances are good that you will find it in the glossary at the end of the book. In addition, a bibliography and resource list will help you find more information as well as people who can help you take control of the energy systems in your life.

For me, each of these pioneer stories has a theme, and I placed the stories among the chapters according to my sense of that theme and according to a larger sequence that begins with reducing your energy dependency in an existing residence and moves forward to consider ways you can design and build a truly independent home, ending up in the final chapters with the larger social and political implications of the solar age, dawning at last.

CHAPTER 1

DECLARING INDEPENDENCE

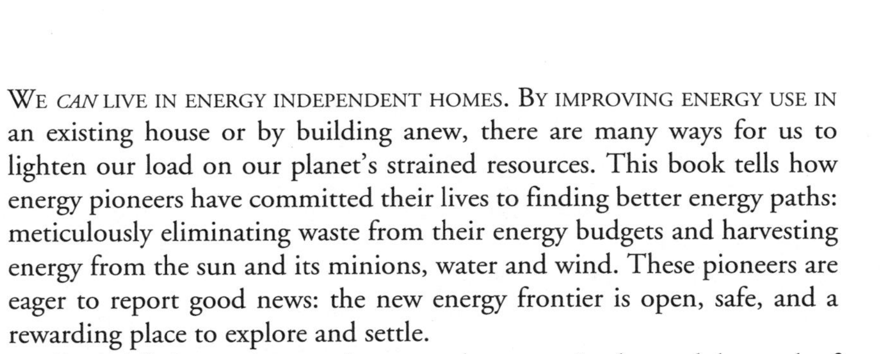

WE *CAN* LIVE IN ENERGY INDEPENDENT HOMES. BY IMPROVING ENERGY USE IN an existing house or by building anew, there are many ways for us to lighten our load on our planet's strained resources. This book tells how energy pioneers have committed their lives to finding better energy paths: meticulously eliminating waste from their energy budgets and harvesting energy from the sun and its minions, water and wind. These pioneers are eager to report good news: the new energy frontier is open, safe, and a rewarding place to explore and settle.

In rapidly increasing numbers, people are moving beyond the reach of power lines and seeking more elemental power sources. Those with limited budgets find that the only land they can afford is unimproved. Retirees move to the end of the road to take up a new and challenging vocation: responsibility for their own needs. On and off the grid, for reasons of conscience and economics, we are cutting our energy consumption and looking for other ways to improve our stewardship of the planet. At times the trail may seem lonely, but in every state, in almost every community, more and more families are following this trail, and taking their rewards in a better and cleaner lifestyle, freedom from the economic tyranny of power monopolies, and an abiding hope that others will join and help save the planet from otherwise certain environmental ruin.

This book contains firsthand accounts by energy pioneers, many of whom gambled years ago that energy independence was possible, though they had to find their way to it across a wilderness of unknowns and loneliness. As pathfinders, they speak now to guide us on a well-marked, increasingly well-travelled way. They bring us good tidings: by following the ever-widening path of energy independence, we may continue to enjoy our present comforts, and without fear that we do so at our children's expense by contributing to the ruin of our fragile globe. The tidings they bear apply everywhere the sun shines, promising a sustainable livelihood can be made available for all the peoples of the world.

On their own personal voyages of discovery and invention, these renewable energy pioneers found romance, intensity, and fulfillment beyond their expectations. By tracking the source of electrical energy back behind the wall outlet where most people stop, they have discovered an intimate and restorative connection with the sun, the earth, the seasons, and the panoply of nature, integrating the best and most modern technology with ageless wisdom about the world. At the same time, they found that shared energy-awareness has united their families and restored the sense of family purpose so often lacking in the plugged-in world. As one who has travelled the path myself, I enjoy the stories these heroic pioneers tell, benefit from their guidance, and gain insight and inspiration from ideas born out of the high and low points of their journeys.

The Dependent Home

The history of homes extends clear back to the beginnings of history itself. Both ideas, home and history, probably occurred to our foremothers while they huddled in a cave, weathering a storm. All forces radiated outward from the cave where the cave-dwellers conducted their lives, where they birthed and died, prepared food, and sheltered in stormy times. Originally these people were wanderers, and caves were the center to which they returned year after year, around which they learned the arts that gave birth to culture. For millennia, we have elaborated that culture, but until quite recently, the home remained a closed system. Energy, food, clothing, and education were generated domestically from locally gathered resources; goods packed from afar were precious, rare, and easily done without.

In the twentieth century, the word troglodyte, another word for cave-dweller, has become a pejorative term for hermit, one who rejects modern life and embraces the oldest, most primitive values. Most humans have proudly left the cave, and have taken up dominion over the world, ruling it for human benefit, taking its resources, bending its other creatures to human use. Accelerating transportation has enabled us to look farther and farther away to satisfy our needs. In the earlier part of the twentieth century lettuce was available only in California and the far southern parts of the United States for most of the year. My father vividly remembers the luxury of the first dandelion green salad in the Colorado spring. Lettuce is now uniformly available in markets across the country, year-round. Before the Great Depression, getting energy from far away was impossible; factory and town sites were chosen first for local availability of energy. In the 1930s, great public works projects and cheap resources made it possible to waste the majority of the energy transported while delivering small amounts to faraway places, never reckoning the true cost. Half a century later, we scarcely notice that our dependence has utterly changed the way we live. But economic realities have changed; once-abundant resources have dwindled because of population growth and heedless waste, and dependence is becoming horribly costly.

A dependent house is a forlorn embassy in the global scheme of ex-

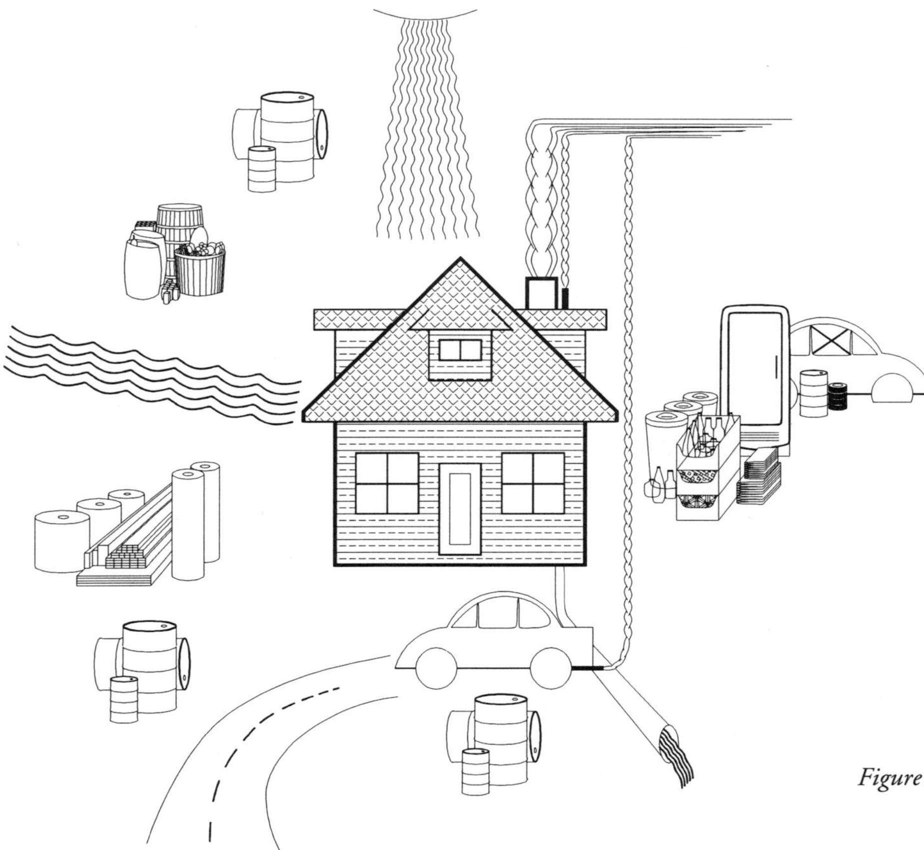

Figure 1. The Dependent Home

ploitation and dominion, cut off from its immediate, ecological surroundings, crowded onto convenient tracts, scarcely more than a place to sleep for its tenants, who are merely consumers utilized by that global scheme in order to perpetuate itself. Dependent householders often have no idea of the origin or true cost of their food, their goods, or their power: at any given moment, the electrons coursing through their walls may have come from Quebec or El Paso, Diablo Canyon or Three Mile Island, generated with oil shipped from the Persian Gulf or coal stripped from Appalachia. Emissions from the power plants will mingle with exhausts of millions of commuting cars to obscure once-clear skies and fall as acid rain.

Today, many of us who live in independent homes, or are moving towards independence in our homesteading, do so because we have concluded that the schemes by which we bring commodities and fuel from far away to satisfy our most basic needs are mindless, abusive, and likely to crash. We seek a happier, sustainable balance between dependence and independence. My life is enriched by chocolate and oranges, neither of which grow well where I live, and both of which I could live without. Energy, on the other hand, is essential in my life, and falls freely from the sky here in Caspar, just as it does in Alaska, Vermont, Mississippi, and Hawaii, wherever the sun shines. Even when sunlight is feeble or absent (as it is north of the Arctic Circle in winter) the atmospheric phenomena the sun stirs, winds and falling water, provide clean and unfailing sources of energy. Through careful reduction and management of my needs, and a comprehensive energy-harvesting strategy, I can produce enough power for all of my household uses—heating, cooling, lighting, pumping, washing, working, playing music—with energy generated in my own yard.

An independent home is a very hopeful political statement, declaring that we believe it is possible to live comfortably and responsibly on the energy that comes to us naturally. What would also be admitted by many who live in independent homes is that we have grown tired of trying to get the system to respond to obvious global pressures, and so have lightened the load by disconnecting from the system, and providing for ourselves. Along the way, we have rediscovered lost virtues.

The Present State of Energy

Before plunging into pioneer stories and hardware, electrons and kilowatts, review with me the present state of energy.

Books, articles, speeches, television specials, movies, and every other known form of communication have documented our dire energy straits. (There is a short list of valuable current literature in the bibliography at the back of the book.) This slow passage, from heedlessness in a time of plenty to careful and attentive use, figures throughout this book as a motivation for independent homesteaders to achieve astonishing transformations and changes in consciousness. To survive, we must devote ourselves to finding solutions. Of the many complicated problems, which do independent homesteaders consider most urgent?

- Our planet is under assault for her mineral riches. Many of these are limited and difficult to extract without damaging life. Our fossil hydrocarbons (flippantly referred to as dead dinosaurs) are quickly becoming the most precious and resistant to retrieval.
- Inattention to what should have been obvious, and short-term thinking including a lagging perception of plenty, has led to wasteful energy habits. While accelerating the depletion of resources, this sloppiness has led to serious pollution.
- We have come to expect a life of full employment and comfortable, wasteful consumption based on the assumptions of perpetual growth and unlimited resources. As these assumptions are called into question, we fear for our comforts and habits. When change is in the air, our first response is to fight it.
- Changes in the way we work, learn, and live have accompanied changes in the way we use energy, to the detriment of values that have long been held important; family, home, work, and nature have all been damaged by precipitous change. Most of us are concerned, but few have good, relevant, constructive ideas about what to do. (Good news at last!) Our wastefulness and superficial attention to the quick energy fix offers a shining hope. We waste more than half our energy, and new technologies are available that allow us to adjust our habits, eliminate waste, and thereby preserve our comforts.
- The best energy sources, the miraculous flow of light from our sun and its natural effects, the planetary flow of water and wind, are relatively untapped and are more available to us now than ever. Using safe, proven technologies, we can harvest renewable energy to sustain our own comforts and share these with all the peoples of the world while preserving our planet's creatures.

Petroleum has literally greased the skids of the industrial age. Oil enabled expansion and technological mastery beyond the wildest dreams of the most far-seeing futurists of a century ago. But the age of easy petroleum is over. We must stop relying on its convenient but dirty energy content, and preserve remaining reserves of petrochemicals for production of durable goods that require its peculiar molecular constitution.

The sun is our ultimate and only original power source. Most terrestrial sources of power are merely translations, in time or form, from this original solar source. For example, fossil fuels are the compressed, fossilized remains of foliage which relied upon and grew toward the sun many millennia ago. Burning plant material or biomass uses the same material without waiting through the millennia; sugar mills burn bagasse, their waste, to generate so much power that they can supply electricity to the mill's company town as well. Hydroelectric power rides the downhill side of the water cycle: water lifted by the sun from the ocean to the clouds, then blown to the mountains and precipitated, drives the turbines. Wind power taps masses of air moved around in the atmosphere by the sun and the seasons. All this power comes from a sun-driven, atmospheric engine. (Nuclear power, which imitates the sun's ability to convert mass to energy, may be a good idea for powering a solar system, but should be done at a comfortable distance from living things, such as ninety-three million miles, and transmitted benignly as light before use.)

I have long been a foe of disfiguring power lines, if only unconsciously. As a child on my family's long summer travels, I lay in the back seat of the car, my head propped at exactly the right height so I could cut the tops off of power poles with the bottom of the car window, as if my vision were a laser gun (which had not been invented yet). I used to imagine the consternation in my wake as people's electric toasters stopped toasting, their electric lights stopped lighting. I was a terrorist, if only in my intent.

I began the research for this book two decades ago, when I concluded that I could and should change my own energy habits. Waste, where I find it, offends me; at first I suspected, and soon confirmed, that my own comfortable lifestyle, much of it learned in my comfortable family home, was in fact destructive and selfish. As my eyes became accustomed to the new ways of seeing, I began working toward more rational use, where I made as much of my own energy as I could and was as careful with imported energy as I could be. I easily found help in some areas: the Arab oil embargo and the recognition that pollution was threatening some of my favorite cities coincided with the arrival in America of small, efficient cars, and I got one. I was planning a house, and so I paid attention to other homes and the home I was then living in, to the features that worked well and to others that were no more than artifacts of empty habit; I planned to replicate the functional and exclude the reflexive. I was blessed at the time by having (along with a natural curiosity and stubbornness) only enough money to do things myself. Friends and family were puzzled that I seemed to be turning away from their comfortable life-paths. I was puzzled, in turn, by their insistence on smoking, on commuting for hours a day, on being so stubbornly resistant to the messages their bodies were sending them. I did not know what was right, but I had a hunch about what was not right for me, and that provided enough direction to start with.

Rediscovering the land: O'Malley Stoumen's story

Many of the stories in this book started when mine did, in the late 1960s and early 1970s, as a "Back to the Land" movement drew the attention of a generation to that estrangement from nature which had resulted from an overly urbanized and technical approach to education. O'Malley Stoumen goes back to the verdant, deeply indented land of southern Humboldt County in the heart of the redwood forest along California's northern coast.

In a way, this is a romance, in that Jonathan Stoumen (whose own voice you'll find in chapter 13) enters O'Malley's story as the heroic architect in her first summer back on the land. Together, they work the land to habitability.

O'Malley Stoumen and the Stoumen children.

I went camping when I was fourteen years old with my family up to the redwoods in northern California. Every summer vacation we went camping, always to a different spot in the state. But I thought that I had never been in such a beautiful spot, and immediately told everyone that I intended to live there when I grew up. Somehow that always stayed with me, and when I was twenty-one I headed for New York City, saying that I was going there to earn a lot of money so that I could buy my land up in the redwoods.

Well, as luck would have it, that is exactly what happened. In 1970, with money in the bank, two friends and I headed up to northern California and deposited ourselves in a real estate office. I can't begin to tell you how naive I was about buying land, but the one recommendation I had was "make sure it has water." So I said to the agent, "I want a fairly large piece of land that is about half meadows and half woods and has a creek on it. It needn't be all flat as I love to hike." (What a dumb thing to say.) He promptly put us in a land cruiser and took us for an hour drive farther and farther into the boonies.

We stopped on the top of a ridge that was just breathtaking and he said, "We'll have to walk from here, there's no road." It was the first day of January and it was sunny and quite cold. This turned out to be good luck for me, because even though I had never heard of a southern exposure, I felt warm whenever we looked at land on the south side, and cold as we looked at land on the north. So I unwittingly ended up with 160 acres of land with a beautiful southern exposure, half woods, half meadows and a half mile of creek (with fish) going through the bottom. Nobody had said anything about gravity flow.

I was so anxious to be there that as soon as the summer came I took my sleeping bag and headed up. It wasn't until then that the size of the project I had undertaken started to hit me. There was nothing there! There was no house, no water, and no one around. I stared thinking that if I was going to build something myself it had to be something easy and preferably out of a book. At first I thought of an "A-frame," but I had never really liked the feel of them and so I decided on a dome. I purchased *The Dome Book* and promptly started making my list of materials and cutting all the two-by-fours into their correct shapes at a friend's house.

I made weekend trips up there with anyone I could talk into going, and started on the foundation (if you could call it that). I picked a terrible spot right out in the middle of a clearing with a little spring running under it. I thought it would be so pretty having the water there. It never occurred to me that after it had rained the expected 120 inches in a winter, the tiny spring would gush out with power to move mountains. We poured tiny pads, and put pier blocks on each one, nailed some redwood posts to them and made the framework to hold the plywood floor. This took several months as none of us knew what we were doing.

And then along came my rescuer. I had met Jon in New York. He had graduated from Cornell architecture school, done some graduate work there, then worked for the City of New York. We had been writing, and he was coming to visit me, and was very intrigued with the idea of a dome. He arrived and the first thing he said was that we would have to tear everything down. He said the house was in the wrong place and stuck out like a sore thumb and that it should be moved over next to the trees where it could still get the warmth of the sun but nestle in with nature. He said it needed a proper foundation and the ground was not stable where the spring was. I finally agreed, but just couldn't take part in the demolition since so much misguided love had gone into the dome. I sat in the woods and sulked for two days while he tore it down. We set up a small semi-permanent camp in the trees, and he said first we had to find water, as it was too far to go to the creek every time. We combed the land and finally found a little spring, built a redwood spring box and piped the water over to our house. We were water self-sufficient! It was a good start.

Jon never had thought of himself as a solar or environmental architect before, but we were forced to live with nature, and she is the best teacher. The first time it really hit him, we had the dome almost all sheathed and we were lying down for an afternoon siesta, trying to let

The Stoumen farm in Humbolt County of Northern California.

some of the heat of the day pass. We had left an opening for this huge south window and we couldn't get away from the heat or the sun because of it. Jon jumped up suddenly and said we could put eaves there that would block out the high summer sun but let the light in during winter. It sounds so simple and sensible and obvious, but when it hits you for the first time it is like magic. So we started on our campaign to have our house work with nature instead of against it, to have it be a collector instead of a consumer.

I don't think we ever even sat down and said that we wanted to be completely self-sufficient; it was just that we live so darned far from town and the conveniences that go with it. We did know that to continue working there we had to earn money, and that Jon wanted to continue his architecture. We also discovered that it was awfully hard to do architecture by kerosene light, and knew that electricity was much cleaner and safer. We had a small windmill that our ingenious friend Jim had built us for a wedding present, but it was very small and we knew that we would need more for those dark winter days of drawing. Jim suggested and ordered for us a 4-kilowatt Electro wind generator from Switzerland. This meant we had to add a room onto our house that would house all the giant batteries. Fifty-six 2-volt batteries! The windmill was a joy to our lives, and made us become more deeply involved with nature. Our lives seemed to revolve around it. When the wind was blowing I would vacuum, make butter with the blender, bring out the power tools and the hair dryer. When it stopped we would conserve. The batteries were powerful and we could go a long time without wind if we were careful. It was also a giant weather vane: when it would turn around and point south we knew to put away all the tools, as it would soon rain. The sound it made was the most pleasant of all. Usually a soft swishing sound as it danced gracefully around. It had its wild sounds, too, when a storm would come up. Before we hooked up the automatic shut-off, there were many trips up the hill in the middle of the night to crank her out of the wind. But those sounds became familiar, too: when to leave it on for maximum power and when to shut it down. Jon somehow still got work out in the middle of nowhere and was dying to try out his new natural ideas. Some work we traded for. One fellow owned a portable sawmill and came and sawed up our own trees into all the lumber we needed to build an enormous barn and architecture office. Jon drew him beautiful plans for his little house in exchange. We got our first real break when Jon entered two of the houses he had designed and built in a HUD-sponsored national competition. He won two first prizes and ten thousand dollars plus a lot of publicity.

As he was taken away more often from home, I found I had fewer and fewer reasons to leave. We had built a beautiful barn, now full of horses, cows, chickens, geese, ducks, turkeys, and so forth. We had planted fifty fruit trees and sixty grape vines and had perennial and annual gardens full of vegetables. The only thing we were really short on was water. There was plenty of it, but it was all down at the bottom.

We had originally dug a small septic tank and had a flush toilet but soon found out what a water waster that was. One day when two other solar-minded architects came to visit, they all sat down under a tree to brainstorm the perfect compost toilet. Jon had always wanted to do a building with a sod roof for a client, but he usually liked to experiment on us first. We also needed a woodshed, so all of this was to be incorporated into the building. One complication was that the best place for it was taken up by a winter-running creek. So these three guys sat there for hours with all these problems and came up with a beautifully simple building that spanned the creek. You had to walk up a few stairs to the toilet but you were rewarded with a beautiful view. We used dry compost from the barn to cover after every use and it all dropped down to a three-tiered system below. The front of the building faced south and the big door to the lower chamber was really a window made from a piece of fiberglass. We called it the solar assist. The sun beating in would really speed up the process. And with the venting there was no smell inside. The finished compost, which was taken out only once or twice a year, was beautiful and sweet smelling. We never used it on our vegetable garden, but it was great for the flowers and fruit trees. All our grey water was piped to various places and we moved the shower outside, too, surrounded by a bamboo screen. For hot water we put up a couple of solar panels and a coil in our woodstove, both leading to a high insulated tank. It was beginning to feel just like downtown.

Jon's work was really picking up and we felt like we needed a phone. There was no such thing as cellular phones then, so we bought a used-model car phone and installed it in our truck. We drove to town and got our number and got it working. Then we headed home and installed it in our home and bought a couple of photovoltaics to power that and a small car radio-tapedeck for music.

Still we had the problem of water. And we had planted an enormous amount of things to be watered. So we started researching drip irrigation. We had heard it was used a lot in Israel but no one around us had ever heard about it. We took a trip to the Bay Area and I started calling everywhere on the phone. We kept getting referred from one place to another until we finally found a commercial flower grower in Half Moon Bay who said he had just been to a dripper convention and showed us a lot of complicated samples that he had brought home. Then he pulled out a very simple design. It looked like a little threaded pipe fitting with a hole drilled in it and a piece of black plastic tape wrapped around it. He said this was the simplest and cheapest of them all and that was what he was going with. You had to have very low pressure (which is what we had) and a filter, but if the drippers did clog you could just increase the pressure a little and the plastic would bal-loon up and clean itself out. He gave us his source and all the information and we ordered a lot of stuff. Jon maintained that the reason they worked so well was confirmed by the fact that every time we had a hole in a garden hose or water pipe, we would try to wrap it with plastic tape and it would always drip. I spent days cutting and gluing pipe but all the components were so easy to handle; the pipe was only quarter-inch and very flexible, so that during the winter I just rolled the sections up on my arm like ropes, tagged them to say where they went, and hung them in the barn. We are still using some of those original drippers even though they are eighteen years old. So absolutely everything went on drippers, things starting thriving, and weeding became less of a problem.

I think my pride and joy in our solar world was my solar oven. I bought a book on solar ovens and made myself one from the plans. It was very crude in some ways. It was made out of plywood with a sheet-metal insulated box inside. It had masonite reflectors with aluminum foil glued on them and a glass door on the front. But boy did it get hot. On really hot days it would almost reach four hundred degrees, and even on cool days it was never below three hundred. So my life was revolving around the sun in another way: if the day was really hot, I knew I could make things that required a hot oven, and on cooler days would put in a slow cooking item. In the summer hardly a day passed that I didn't use it. I made all our bread in it and just about everything else you could think of. If it was a beautiful warm day and the oven was empty, I would start to feel guilty like I was wasting energy.

Jon Stoumen

But the big problem, and a universal one for sure, always came back to water. One day when walking down by the creek we started talking about all the water down there and how we could get it up to the house area. It was four hundred feet up and a half mile away. We knew we didn't want to install a big diesel pump down below that we would constantly be fueling. We measured the amount of water going past just to see what we were talking about. Armed with that information, we wrote to a hydraulic ram company back east, the start of a lot of correspondence and the eventual purchase of a ram pump. It was a project that took up almost the whole summer and was mostly just grunt work. We carried 110-pound sections of pipe up this steep creek to be put together with pipe wrenches almost as big as I was. We had a big nineteen-thousand-gallon ferroconcrete tank built up above the house, and had a smaller tank down below to catch the water that would shoot fifty feet straight down to run the ram. We built a concrete house for the ram and installed it and then put a big pipe from there back down to the creek where the water used for the power would be returned to the creek. At the highest place on the creek we built a small dam, which we could slip the boards out of in the winter and put them back in for next year. It was incredible sweat and toil, but when we finally got it going it was so awesome and unbelievable to see the water pouring into the tank that far from the creek.

There were lots of other things I did with my time. We had sheep, so I carded and spun and made warm hats and sweaters for the family. I took a four-day course at the local college on cheese making, and since we had two milk cows I was soon pressing, waxing, and storing cheese. Of course there was always canning, drying, and putting up food. We had become almost self-sufficient and that meant putting away for the times that were leaner or out of season.

Suddenly we were confronted with a problem that we had never really considered in our plans. It was time for the kids to start school. The school down the hill would require four hours of driving a day, and the reason we were doing all this was so that we wouldn't have to drive down the hill all the time. Another problem that was bothering us was the dope growing. We were probably the only people on the whole mountain that weren't growing it and we worried about the kids again. For all of their friends it was a perfectly acceptable thing and even part of family life. They all helped in the gardens, sharing the harvesting, the cleaning, the profits, and the smoking. It didn't seem like what we had in mind. I hadn't ever thought of home schooling, or that might have been an option.

So we rented our farm and bought an old house in the little town of Healdsburg. But we still have our farm and it is still waiting for us to return, which I'm sure we will do. Meanwhile the whole experience there is carried with us every day. Jon's architecture was deeply affected, and he will always be doing environmentally friendly work. We still use drippers and conserve water in every aspect. When we remodeled our in-town house, we put in a separate line for all the grey water, and used all natural materials. We use compact fluorescents and low-E windows. I think it is probably more of a challenge trying to accomplish all of this in a city than it is in the country, where you may have no other choice. So that is the challenge we are working on now until we can get back closer to the land.

Looked Like Movement to Me

O'Malley's story may seem to end in a retreat from country life, but she assures us that her return to the city is temporary, for the sake of her children and her husband's work. She has given up neither the farm nor her dream of returning there, and I believe she will return. I expect her to work out an up-to-date version of an age-old pattern of town house and family farm. The farm in the Eel River's basin and the town house a watershed south, along the Russian River, are both within the coastal redwood bioregion; on horseback it would be a wonderful two or three day ride between them.

We can also be sure that those who move back to the city after a time at the end of the road have been changed utterly by the country experience. Their new ways of seeing, of responding to challenges and caring for the land, alter and enrich their responses to the needs of home. Many of the storytellers in this book truly found their voices in the late 1960s and early 1970s. At that time, my decisions and those of many of my generation looked to our elders like a movement, a political reaction; neither we nor they could really say what we were reacting to. The Vietnam experience muddied waters already turbid with civil rights and freedom of speech issues, and issues like energy and waste were forgotten elements in our lifestream. Even though stewardship permeated every current and backwater of my life, at the time this concept was not strong enough to catch and hold my attention. Bullheadedness and myopic self-reliance led me to make many mistakes. I have come to identify a pattern in my responses: (1) I would encounter a stupidity, an inflexibility in the building codes, for example—"You can't let your grey water just run out into your garden, it's got bad chemicals and germs in it." Excuse me, but how could that be? If I am careful to put only good things, biodegradable soaps, toothpaste, food particles, in one end, how do bad things get in there? After all, I did the plumbing, and knew there were no secret trap doors to the evil empire. "I don't know," says the sanitarian, "that's what the code says." So what happens if I ignore the code? "Maybe, you get sick. And I shut you down." What a challenge for a political malcontent like me! (2) Generalizing from this, I would decide that the whole code was stupid and all inspectors inflexible and powerless. My father calls this throwing out the baby with the bathwater.

Reinventing the Cave

It is very time-consuming, re-evaluating all the assumptions back to the beginning of human habitation, reinventing shelter, but that is what I tried to do. And I felt isolated, like the Lone Ranger, asking questions which felt basic to me and being assured by an authority that it was best not to ask but simply to accept the fashions and standards then in place. I suspect that I, and most of those stumbling about in the dark looking for better ways, unconsciously sought loneliness. Travelling to complete this book, I finally saw that I was not the only one who chose this demanding and eclectic path.

For me and for many, cost was often the primary motivation. The core of my new house grew slowly, and so walls that would one day be inside the house survived battering winters exposed to the cold and the rain. In this abrasive environment, I quickly learned which of my reinvented building techniques held up, and which failed. My waterproofing scheme, using methods recommended in books but meant for more clement climates, was repeatedly vanquished by the aggressive winds and waters of my coastal site, and repeatedly failed . . . and continues to fail, though not so completely, as I write. Nor am I alone: the best architects and builders and the most modern materials and techniques fail under the assault of horizontal rain and a corrosive marine environment. We are trying to extend the margins of habitation, and it will take time and repeated tries to find proper materials and techniques. Meanwhile my home is, and will always be, an experimental shelter. Almost before the first room was habitable, and ever since, I have known how to do it better next time. On my travels, I have discerned two styles: my own, which I call organic, wherein much remains unfinished until inspiration comes and I can envision the completed detail, and a more architectural approach wherein the vision appears to have been elaborated from the broadest strokes down to the smallest details before the first spadeful of earth was turned.

One lesson I have learned well is that initial cost is only a part of the whole cost. In Caspar, water is the most cunningly invasive element. As I struggled to keep water outside my walls, I learned that I had omitted a key ingredient, the impermeable membrane, which should have been at the conceptual and mechanical core of my wall. Building out from this core would have been simple; adding it after the fact was at best difficult, and at worst, ineffectual. No amount of money, after the fact, could fix that error of omission. I will have to rip out whole walls and rebuild them right, and the whole historical expense of the resulting wall beggars the cost I avoided at the beginning. Lesson: If you cannot find time or money to build it right, how will you be able to build it over? So do it right the first time.

Energy-consuming products, from lightbulbs and clock radios to refrigerators and automobiles, usually cost more to operate than they do to buy, and so purchase price is a terrible yardstick. Suppose you are an apartment dweller whose electricity payments are included in your monthly rent. Would it be foolish to buy a light source that will outlast your intended stay? Here, we begin to see hidden social costs. If you can demonstrate to the landlord a pattern of reduced electrical costs, might he not offer a rent reduction? Might you be wiser to separate utilities from rent, so that any economies might directly accrue to the economizer? The landlord stands, in this case, for society: by burning fewer kilowatts to generate the light we need, thereby reducing energy consumption, slowing the race to exploit energy sources, and decreasing emissions associated with their extraction and incineration, we benefit humanity. Even though a cost may be concealed, it is still real, and somewhere, someone pays. There is still no free lunch.

What hidden costs are associated with light sources? The costs of creation and disposal, which are often hard to know, are external in the sense that they are not entirely counted in the price we pay. The tungsten, tin, and lead are mined, sometimes in earth-marring ways; tungsten is a strategic metal, which means we might have to go to war to protect our source of lightbulb filaments. Compact fluorescent (CF) bulbs also use rare, strategic, and extracted materials. Extraction costs, human and environmental costs, and the cost of continued readiness in case of warfare, allocated for the billions of lightbulbs manufactured, obviously adds a small but undeniable amount to the cost of each bulb we buy, with incandescents

Whole-Life Cost Analysis

As the distance from original source and manufacturer and consumer has stretched, some of the real costs have disappeared into the infrastructure of modern life. In some cases, these referred costs are greater than what we pay for an item; for an extreme example, how much would the people of the Ukraine add to the price of their electricity to pay for the clean-up of Chernobyl?

There are at least two factors in the whole cost of any item: the cost to buy it (equipment cost), and the cost to maintain it (operating cost). To those costs, we should add two more: the true cost to create it, and the true cost to dispose of it at the end of its useful life. Both of these are usually borne by the society at large in hidden ways. An example would illuminate the idea.

Start with a simple whole-life cost comparison: in this corner, the champion and most-favored light source, the incandescent bulb. In this corner, the challenger, a first generation compact fluorescent (CF) light. Considering only the two standard economic factors, equipment cost and operating cost, the champion wins the first round easily: who would buy 60 watts of light for twenty-five dollars when you can buy a package of four champs for under three dollars? But wait! the cheaper ones only last a ninth as long, and more than two packages of four are required to compare fairly, so the challenger appears to be only four times as expensive. If you budget only seven dollars in your pocket to buy light with, you need study no more, and initial equipment cost decides the contest. Wait again! The CF package bears a printed assertion: this newfangled light source consumes only a quarter of the electricity drawn by the conventional bulb. You pay for electricity, do you not? Over the whole life of the two bulbs, which is cheaper? Assuming that electricity and lightbulbs will not increase in price—an exceedingly unlikely assumption!—the incandescent bulbs (it will take about nine) will use so much more energy over the lifespan of the compact fluorescent that the champion's overall equipment plus operating cost will be almost twice (188 percent) the challenger's. Based on recent energy trends, a 10 percent annual increase in equipment and operating cost is reasonable, and the champ suffers even more, costing almost two and a half times (236 percent) as much.

Compact flourescent lightbulbs.

or compact fluorescents roughly equivalent. At disposal time, again, both bulbs are roughly equal, largely recyclable (and universally unrecycled), with one surprising gotcha! marked against the compact fluorescent: the bulb used for this whole-life cost comparison was a first generation, core/coil-ballasted lamp that contains a tiny amount of radioactive material, americium, which was found necessary to get the early CF bulbs to start. No one knows for sure, but we have reason to believe that, were the bulb disposed of carelessly and the americium somehow ingested, it would most likely prove to be teratogenic or carcinogenic, a serious source of sickness (though it has been suggested that one would have to ingest several dozen lightbulbs to be at risk). Can an actual dollars-and-cents cost be assigned to such a risk, or, as the utility industry expresses it, can this cost be monetized?

Newer compact fluorescent lights, which use electronic ballasts, have eliminated the radioactivity and are an altogether improved product. They start quickly, and the life expectancy of the ballast is better matched to that of the bulb.

Thrifty technology: Fred Rassman's story

Once a person is in the habit of buying goods based on the whole-life cost of each purchase, many assumptions about thrift come up for reexamination. Like reinventing shelter, this can become a consuming pursuit, with gratifying results.

Fred Rassman lives with his family in an off-the-grid A-frame in the Genesee Valley of northwestern New York State. He sought a life where he was responsible for the forces that acted on his life and his family. As you will see in his story, this trained engineer wants to know the true cost of anything he does.

I bought the land, ten acres for $1500 dollars, in 1970. I guess I got it because it was pretty swampy then. I built my A-frame on the best spot without thinking about solar or anything: it was just going to be a summer cabin, and I was twenty-one, so what did I know? When I decided I was going to stay, I went to the power company, and they wanted $10,000 to bring power in, so I went without. Didn't make sense, spending ten times the value of the land.

I grew up in Pleasantville, down near Brooklyn, and we had a few trees and a pond or two, but it was city life. My old man took us camping for two weeks every year, but I always wondered, why do this two weeks a year when you can do it all year long? So after I studied engineering at Alfred University, I decided I never wanted to go back to

The Rassman family at play beside their power plant.

the city. Most of the year, I'm director of quality control, and run three labs, for a local construction company; the rest of the year, I take care of my homestead, and now I've got a family, too. Sometimes I wish I'd started the family earlier—I'm forty-four now, and I've got a boy aged five and a girl aged two, and I can't imagine that part being any better.

In 1978 I came across an old Jacobs windmill, and built a tower for it. I'd never seen anybody else's towers, so I built what I thought would work; turned out, it's much stronger than it needs to be. I just took the Jacobs down two years ago. It was still running fine, but, with twenty-four panels on two twelve-panel trackers, I was getting more electricity than I knew what to do with, and wanted to save the Jacobs from wear. I've helped a lot of friends put in Jacobs-type mills, done it in all kinds of weather, ten below and windy, so I'm on borrowed time now, and I'd just as soon stay down off the tower and watch the trackers follow the sun.

Anyway, in 1978 I got electricity, and started to make things a little homier. I just got married six years ago. I guess Mama checked me and my place out pretty close, and she's pretty happy out here. We live alone in the woods, and we love it.

I'm a hands-on kind of guy, I work on my own cars, and I've built and maintain everything here: the house, two sheds, and a pole barn. We garden, more as a hobby, and have friends who have a roadside stand. I raised pigs for awhile, and it was satisfying knowing it was my meat I was eating, but I needed the space, and the feed got too expensive. Like I said, watching the sun shine is more fun. I've added two wings to the A-frame, so now we've got about sixteen hundred square feet. Up in the peak I've got a batch solar water heater, so I preheat my well water, then I've got two super-insulated water heaters that run off my excess solar electricity, a twenty-gallon for showers that runs about 160 degrees even on a rainy day, and a forty-gallon for the kitchen, that runs a little cooler. Right out the big octagon window I put in a half-acre pond, for swimming after work in summer, ice-skating in winter. I've reclaimed the forest too, in pines, and it's too pretty to cut, so I pick up scrap wood at work and buy the firewood I need. The house is grandfathered, and I guess the county doesn't know I added on. We've got a composting toilet too, just because I hate to waste anything, you know?

When I need to put something together, it's easy enough to get a book and read how to do it. Then I can pretty much make it work. I've got my PVs at 38 volts, and I have six tons of telephone batteries set up in several banks, with big old Frankenstein knife switches to control them. In the house, we've got 32 volts and 12 volts DC and 110 volts AC. I like to keep everything I can—TV, lighting—on DC. The only real sunny spot is two hundred fifty feet from the house, so I ran some big copper wire down there, and still lose a couple of volts, but it doesn't matter so much at the higher voltage. The batteries are the best part of the system: I got them from the phone company back when they'd just let you haul them away or charge you for scrap. I had a contact at Western Electric, and hauled some batteries in from Indiana, some from Pennsylvania. Some of them are thirty years old, but they're in great shape. I started out with golf cart batteries, and I guess it's true that you've got to ruin one set to learn how to use them, because those only lasted me a year. With a wind generator, it's harder to get that last volt.

I guess I could get grid power easily now—it's just a quarter mile away. But what a great feeling, making my own power. If you've never done it, you just don't know. We get ice storms here, and folks around us lose power for several days, and have no water, and can't cook, while we never have any problem. And the things they've got to do to generate electricity, nuclear and what-not, make you sick. Here in Allegheny County, I guess we're the poorest county in the state, and so they're trying to dump their garbage and their nuclear waste on us, and it makes you think.

I just like knowing this is all mine.

Describing Electricity

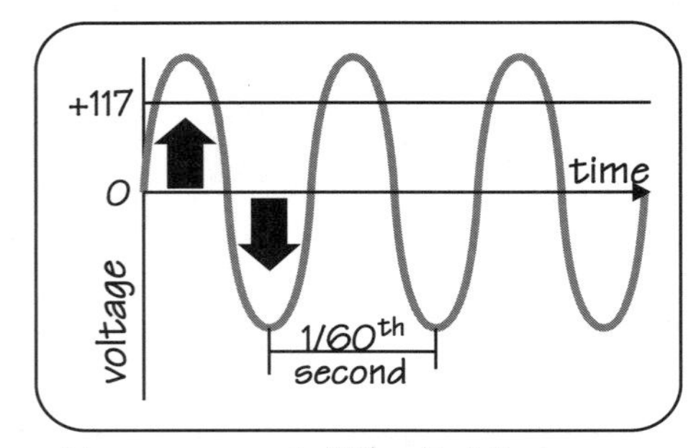

Figure 2. House current is electricity delivered at 117 volts as sine wave alternating current. The electrons flow back and forth through the wires, changing direction twice every sixtieth of a second.

In all but the simplest of homes, electricity is our energy source of choice. There is a wholesomeness and purity about electricity that appeals to us, and its intention and intenseness makes us happy to surround ourselves with it. It is invisible, undeniably modern, a pervasive and reliable energy source that reaches into every corner of our homes.

Electricity is a flow of electrons, but to describe that flow we require units more our own size; we cannot count electrons any more easily than we can describe the flow of water in a pipe by counting the water molecules.

We routinely work with electricity delivered to us in two ways: direct current (DC), the kind of electricity that comes out of batteries, flows like water in a pipe, in one direction; alternating current (AC), our common house current, flows back and forth through the wire, changing its direction one hundred twenty times per second.

Direct current (DC) is easy to understand. It is sometimes likened to water flowing in a pipe: you open the tap (flip the switch) and out flows water (electricity) at a flow rate (current) determined by the distribution system in gallons per minute (amps) at a given pressure (volts). Direct current electrical devices are using the flow itself, while AC devices often use the changes in the flow.

Sources produce electricity. Batteries and outlets seem to produce electricity, but they are simply agents: batteries store energy, and the electrons flowing from an outlet may be have been pushed by sources anywhere on the North American continent.

Loads consume energy. Anything that uses electricity to perform work, from a battery-powered wristwatch to the winch that lifts a drawbridge, is a load on some energy source while it is in operation.

The amount of electricity that sources produce and loads consume is measured in watts; a kilowatt (abbreviated kW) is one thousand watts, and a gigawatt (gW) is a billion watts.

In the United States, grid-supplied house current is a complicated electron flow: alternating current somewhere between 110 to 120 volts. The voltage alternates sinusoidally sixty times per second between two peak voltages, somewhere around plus and minus 167 volts. This means that the voltage at any given instant can be measured as +167 volts, -167 volts, or anywhere in between. The nominal voltage is the root mean square (RMS), or the effective average received by a load. On an oscilloscope, a device which depicts electrical changes over time, house current produces the smooth sine wave shown in the diagram (see fig. 2). Electricians like to talk about how loads "see" the energy: where most of them see the RMS voltage, a few, including early core/coil ballasted compact fluorescents and laser printers, want to see a peak voltage up in the mid-160s or they won't work.

By comparing the energy used to generate electricity with the work performed by the electrical device it drives, we may calculate the efficiency of the process. Some tasks can only be accomplished with electrical power—most electronic devices fall into this group—while others are extremely well suited to electrical power, and still others use electricity very inefficiently. For example, heating anything electrically is inefficient, because only the changes in electron flow are used, not the flow itself. We overlook this inefficiency because electrical heating is so convenient. It is hard to imagine a propane-fired fuser in a copying machine or laser printer, even though the fuser, which melts the colored plastic onto the paper, is the single most wasteful part of these wasteful machines. Electric motors can be made to use either AC or DC, but a DC motor is more efficient because it will consume less energy than an AC motor to produce a given amount of rotational power. Most electronic devices, which neither heat nor whirl—clocks, televisions, computers, anything rechargeable—convert all or part of the alternating current they draw to direct current before passing it along to the parts that work for us, and this conversion is inefficient; we can experience that directly by checking the heat emanating from a device (for example, a television set) that is not really meant to be a space heater. The best uses of electricity are the efficient uses.

Alternating current (AC) was invented for convenient distribution over long distances. Electricity can be generated either as direct or alternating current, depending on the configuration of magnets and coils in the generator. Alternating current has nothing to recommend it except that electricity packaged that way moves more manageably over great distances. On the home level, except for wasteful inductive loads (heating, transforming) it is not particularly convenient to have alternating current. For most of the best uses of electricity, house current's alternations are a nuisance which must be converted by each appliance to a more suitable form, usually wastefully, and most appliances do this conversion as part of their power supply. That is what those energy criminals, the phantom loads, are doing underneath the sofa (see chap. 2). Anyone who has tried to get the buzz out of a stereo or telephone set can testify to the nuisance of the pervasive sixty cycle hum. We are also becoming aware of possible health risks associated with constantly bathing ourselves in this electrical pollution, called Electro-Magnetic Radiation.

The Hidden Costs of Electricity

In terms of hidden costs, those associated with the kilowatts we pump through our lights are by far the most expensive and best hidden. Let us delve only briefly here, identifying the issues without trying to quantify.

- *Costs of extraction and transportation.* These are considerable, and largely hidden. Most power generation presently uses natural gas and coal. Both resources are still reasonably abundant on our continent, and supplies should last well into the next century. Since both are extracted, their true cost differs from what we pay because the actual materials are given up freely by the earth. Actual costs include the profit taken by the resource holder, and the monetized costs of extraction and transportation from mine or wellhead to the power station. Hidden costs, called externalities because they are external to the cost calculations used by the power industry, are found all along the way: health risks and damage to the environment from extraction to delivery, emissions and losses during transportation, and (hardest of all to put a price on) the possible extinction of lifeforms or materials that may turn out to be irreplaceable for some future, crucial need.
- *Inefficiencies in generation and distribution.* No one knows for sure where the electricity lost during transmission goes, but it can amount to half the total generated. Costs, in terms of deformed calves, decreased dairy production, and other effects of stray electricity are well-documented, and quietly dismissed by utility companies where they are shown to be undeniable. Electromagnetic effects of high-tension transmission lines on humans, animals, and plants are unknown. Can we even begin to evaluate the harm done to our magnificent scenery by powerlines draped insouciantly over hill and dale?
- *Waste created when electricity is generated.* Conventional power plants create megatons of fly ash and scrubbed sulfur; belch more tons of noxious fumes (although, as dead-dinosaur burners go, they are quite clean); and are, at the end of their relatively short lifespans, impressive stacks of used, useless equipment. Heretofore, the standard utility response to this problem has been to bury or dump the pollutants and let the machinery rust in place. In the extreme case, nuclear power plants bring a whole new meaning and time frame to the costs of disposal. Decommissioning a nuclear plant requires that many of its parts and ingredients be encapsulated for periods up to twenty-two thousand years at the very least. This is,

> above all, a building and communications challenge far beyond anything ever undertaken: our oldest structures are barely a tenth that old, and how can we be sure that our warnings— "Keep out! Radioactive waste!"—will be intelligible in twenty centuries? No scientist denies that a nuclear plant's whole cost of decommissioning and abatement, though never included in the new plant's cost/benefit analysis, will most likely exceed its construction cost a thousand fold.

Those who study externalities, and who work to reduce our irrational energy consumption by helping us pay the whole-life cost of energy, conclude that electricity's true cost is easily twice what we now pay. From a global perspective, energy now costs more than the planet can afford.

In our own homes, we could use half as much electricity as we do now, and yet the impact on the planet would scarcely be reduced. We must go further, and institute far more sweeping changes, so our individual efforts will compound themselves rapidly—before the petrochemicals are used up and our environment becomes inhospitable to all but the insects and genetically engineered microbes that love to eat toxins.

CHAPTER 2

HOMEWORK: *Outgrowing Dependence*

DEEP THROAT, THE ALL-KNOWING WATERGATE INFORMANT, ADVISED US TO follow the money. Good advice for solving any puzzle with money in it—the prescription that lightens the weight of our personal contribution to the planetary energy overload follows quite closely the money we spend on energy for our homes. We know that energy costs will rise sharply in our lifetimes. The primary energy consumers in a house are space heating and cooling, hot water, refrigeration, lighting, cooking, and random plug loads: entertainment and office devices, an infinitude of small appliances that load the electrical system in a modern home. By conscientiously examining every energy use, we will find ways to lighten our load and simplify our lives.

In the year 2010, a decade into the next millennium, 80 percent of the houses that presently exist will still be standing, and many of us hope to be living in them. Considering the irreplaceable energy already bought and incorporated into this housing stock, and the generations of emotional investment in these old buildings, in many cases losing them would be unconscionable. Nevertheless, most of these houses are energy pits into which we throw money, to say little of the environmental future of our race and that of thousands of other species we will take down with us if we do not wise up.

We are a nation of tinkerers and do-it-yourselfers, and our homes are often our favorite projects. This is especially true for the independent homesteader: the home becomes the *magnum opus*, the great work. Independence requires dedication and discipline. We already know that constant vigilance is the price of freedom; this chapter tells what we must know and do to secure our energy independence.

Cost-Effective Home Energy Systems

Many of our homes, as well as the major appliances in them, were designed and manufactured before public energy awareness dawned, or during the time it took our massive manufacturing apparatus to reform its engineering and production practices and use of materials. The energy industry, responding to escalating resource costs, awoke first. Other in-

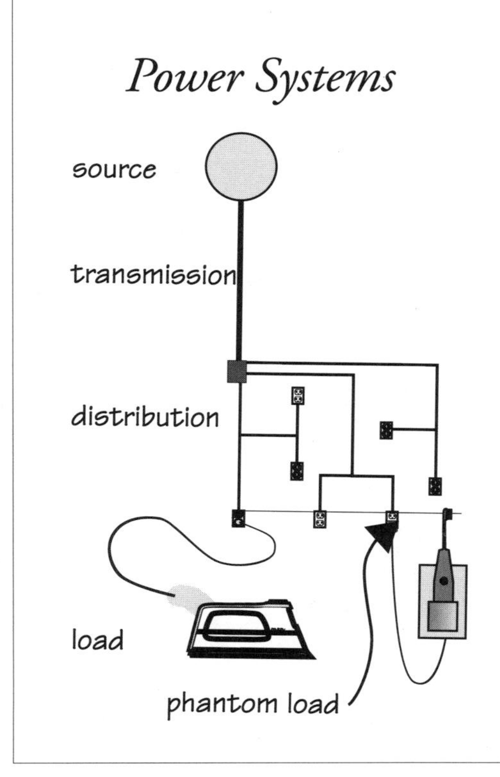

Power systems are composed of four parts: source (where the flow of energy originates), distribution (the means of moving energy from the source), controls (which can measure, direct, and control the flow), and loads (which convert the energy into power). These words are most often used to describe electrical systems, but they apply equally well to any energy system. For example, a woodburning stove is the load at the end of an energy chain that starts in the woodlot, where trees are the source.

Electricity is a flow of electrons that is generated at a source, moved and controlled with conductors and switches, and used by loads. Electricity can be stored in a battery.

Figure 3. Electrical system components.

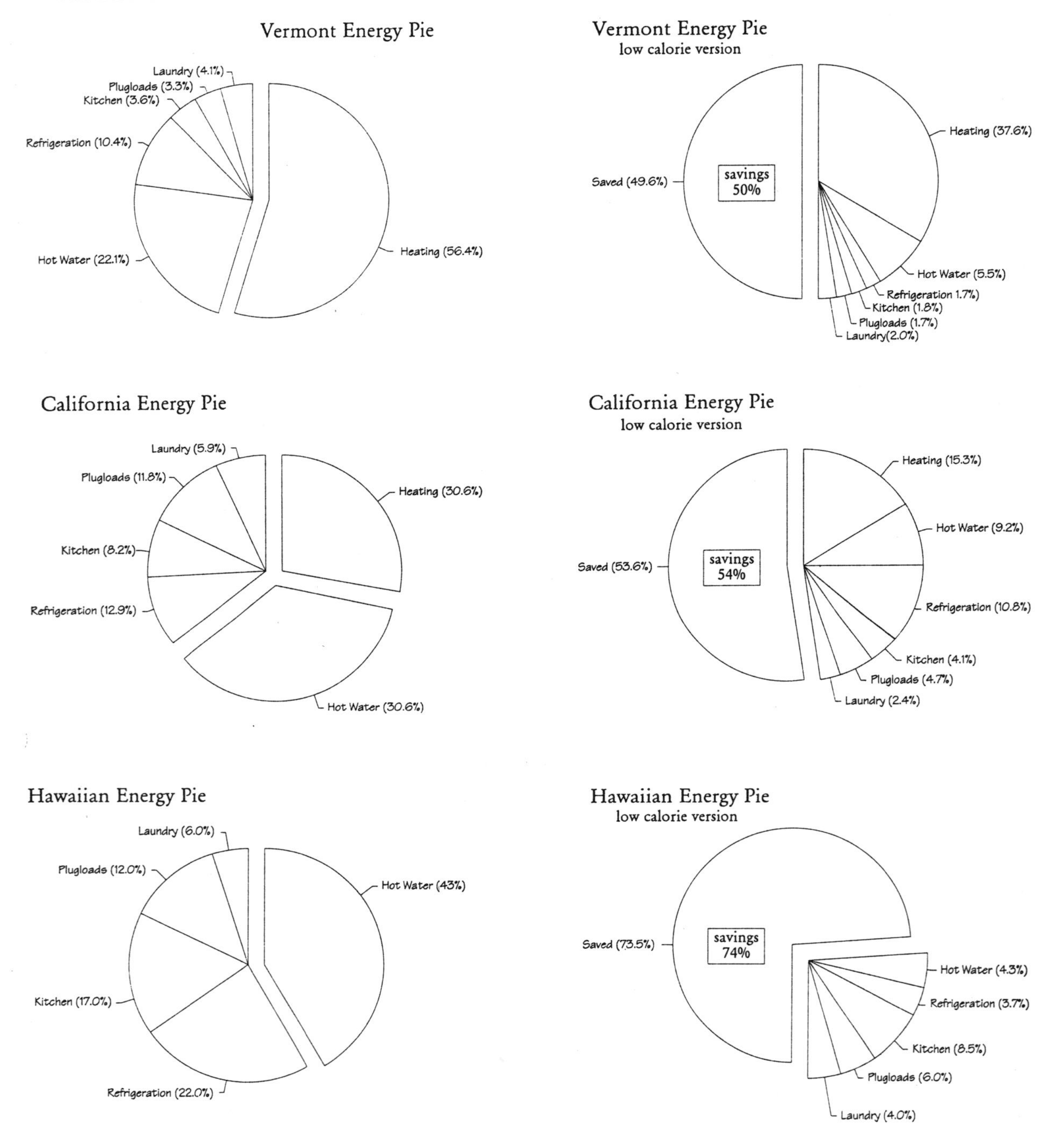

Figure 4. Normal and low-cal home energy consumption in Vermont, California, and Hawaii.

dustries are responding in order of the depth of their entrenchment in, and connection with, energy and pollution. This new awareness has reached, but not yet reformed, the automobile industry. Homebuilding, bound as it is by fashions and building codes, will typically be last to change, unless alternative architects and builders work quickly for reform from within.

To attain cost effectiveness in our home energy system, we must apply measures that repay their added cost within a component's effective life, thereby reducing our energy dependence. This standard, called *payback*, can be calculated and applied quite rigorously if one is of an accounting turn of mind; an example of the necessary calculations will be encountered later in this chapter, in our discussion of refrigeration. In calculating payback periods, I have found it useful to do so twice, once assuming straight-line energy costs, where the amount paid per kilowatt hour or therm does not increase from present levels (unlikely but very conservative), and a compound increase model of 10 percent per year (which is, I hope, the worst case). The real payback should be somewhere in between (see fig. 5). Cost of capital is left out of these calculations, and this cost would be substantial, because in almost every case we will have to buy more expensive equipment, expecting to take our payback in reduced energy costs over time. Let us say, simply, that we will forego interest payments in partial restitution for past energy depredations.

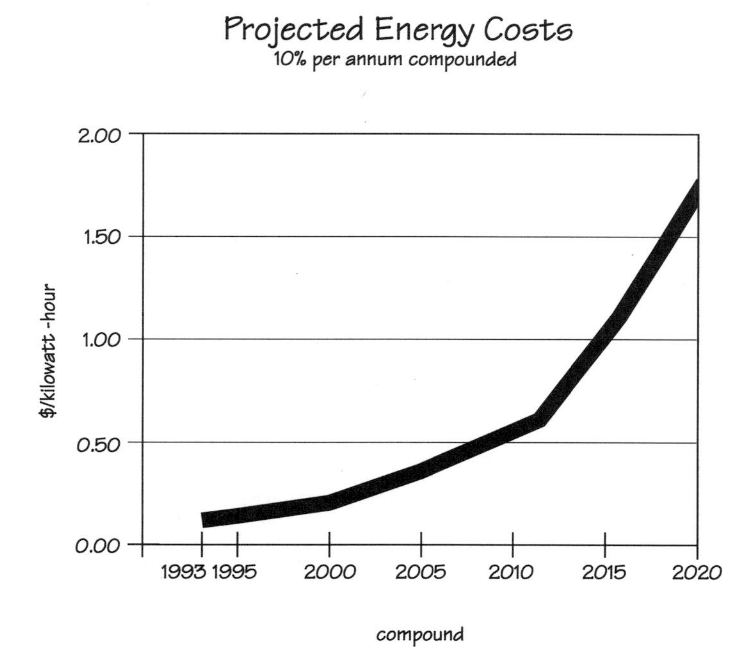

Figure 5. My pessimistic prediction of electricity cost over the next two decades assumes a 10 percent annual increase.

The basic principle of payback is nothing more than common sense, belatedly applied to an area where we enjoyed the brief and probably foolish luxury of ignorance brought on by a surfeit of cheap energy. In 1960 we didn't know any better. Now we do.

Retrofitting: Steven Strong's story

Steven Strong designs solar electric systems and environmentally responsive buildings with Solar Design Associates, an architectural and engineering firm in Harvard, Massachusetts. Here are some strong ideas about bringing energy efficiency home.

Retrofits are no architect's passion, but we must face the fact that the majority of the buildings for use in the next generation are already built. It's sometimes difficult and often unglamorous work but there are far too many existing buildings to ignore. What can we do with these structures? After increasing their thermal integrity and optimizing their energy efficiency, we must retrofit them to produce some or all of their own energy on-site from renewable resources.

It turns out to be reasonably easy to address these problems, and most of the solutions are equally appropriate to new construction. The basic hierarchy of options can be prioritized in terms of cost-effectiveness and return on investment.

The first priority is thermal integrity. Start with the building envelope. Upgrade the insulation and gaskets, and eliminate infiltration. Consider controlled, heat-recovery ventilation. Check out the windows. Regular double glazing has an R-value of one and seven-tenths; today's high-tech window's R-value is nine.

It is simply inexcusable not to employ the very best windows you can buy. Just in terms of resource allocation: microns-thick, low-emissivity (low-E) surface coatings and inert-gas fills justify themselves. Glass itself will last more than the lifetime of the building and, with the investment of less than an ounce of additional high-tech materials, you can cut energy waste by a factor of two, three, or more over the building's entire life.

In the last three years there has been a full-scale revolution in the windows you can buy. For example, Andersen no longer sells standard windows. When you buy any Andersen product today, you get argon-filled, low-emissivity windows whether you ask for them or not. This is a major, major advance. By effectively doubling the R-value of their entire product line, Andersen has, almost on its own—because of its volume—

Steven Strong's solar subdivision in Gardner, Massachusetts.

set a new industry-wide standard for energy efficiency. Other manufacturers have then followed suit.

As the race to develop "super windows" continues, new technologies recently introduced provide almost three times the R-value of the standard low-E Andersen units which are themselves about twice as good as ordinary "standard" double glazing. Let me sound like Amory Lovins for a moment: If we retrofitted the glass in all existing buildings in this country using today's best-available technology, we would save more energy than flows from the Alaskan Pipeline. That may sound like a formidable task, but it's cheaper and easier and "nicer in every way" than another war over oil.

Next, install a state-of-the art heating plant. Many state-of-the-art plants burn conventional fuels at close to 95 percent efficiency, and easily pay back their cost even when they're used just half the year.

Then, look at your lighting, appliances, and plug loads. Shopping for the most energy-efficient appliances is a critical (and simple) thing each of us can do to move the country toward greater energy efficiency. It may even make sense to change our appliances before the end of their design life if you gain significant efficiency. It's fairly easy to figure. Amory calculated this years ago and the utilities are just now catching on.

It's technically possible to build a refrigerator that uses one tenth the energy of the average unit out there in American households. There's an energy revolution in refrigerators on the way as evidenced by PG&E's "Golden Carrot" program in the U.S. [See page 39 for more about the Golden Carrot program.] Refrigerator energy consumption will likely fall by at least a factor of two or three over the next ten years. Other major appliances are also undergoing similar redesign. Finally, after you've tightened up the building and reduced the loads, look at the possibility of producing your own energy on-site. Solar hot water heating is very effective: it's off-the-shelf technology, well-debugged, supported by a good service infrastructure; it's easily interfaceable with existing systems and works equally well in retrofits and new construction.

A home in Steven Strong's solar subdivision.

Further up the return-on-investment curve, you'll find solar thermal systems for combined space heating and domestic hot water, and, finally, on-site generation of electricity with wind or hydro power, where available, and photovoltaics anywhere. It is possible to achieve energy independence with today's technology even in the northern tier of states, where degree-days are high and sunlight resources are modest.

Like high-tech glass, the ultimate solutions will be invisible, with more benefit for less money, in the long run, than we would expect . . . and you won't even see the difference. These changes are coming about, first, because of the energy market, but second, because it is not that hard to do much better. If we, as a society, invest our talents and resources wisely to develop the right set of technologies, austerity and deprivation will simply not be necessary. On the other hand, if we fail to do this, they most certainly will be.

For new buildings, which is what all architects like to do, we must make sure our philosophy is in tune with what we're trying to accomplish on the global scale. In our firm, we will only work with clients who have a high degree of concern for environmental issues. At the same time, our clients want a high-quality living environment. Our houses reconcile this conflict by making it possible to live lightly on the earth without taking a vow of deprivation to do so. Many of these houses are energy-independent and only a few might be considered "inexpensive." That's a difficult issue for me, that our work is looked at as just for the rich. But, it's a fact of life in the profession—whether they want sustainable buildings or not, only people with reasonable means hire architects.

I'm in love with photovoltaics. It is the most compelling technology I have ever encountered—silent, nonpolluting, self-renewing—simply elegant. We powered our first building with PVs in 1979. It was an eleven-story mid-rise and cost a truckload because it was twenty-five or thirty years ahead of its time. We designed and built the Carlisle House in 1980 and the PV system for it also cost a truckload. When I'm criticized on the cost of our PV-powered houses and buildings (usually by the press), I point out the twentyfold reduction in the cost of pocket calculators in the last fifteen years.

It costs real money to bring a new technology from concept to a mature commercial product and early demonstrations are going to be expensive. If the calculator industry had been shamed into giving up in its infancy by complaints that their simple four-function model was "too pricey!" we'd all still be using those clunky, mechanical adding machines. Predicting the future is like taking a snapshot through a keyhole, but I feel our clients are beginning to invest in photovoltaics because they see the watershed of change coming. We've got the technical solutions now; all we need is the political will to implement them.

I realized many years ago that we simply can't change the world, one custom house at a time—they may be fun to do but there just aren't enough to make any difference to society. So we became politically active, working with state and local policy makers and utilities to find ways to restructure capital resources and introduce energy efficiency and renewables. Our private clients have supported our explorations of solutions that, in the near future, can then be shared with working-class homeowners who simply can't afford to be early innovators.

Off-the-gridders are true pioneers who have created a significant base marketplace and proving-ground PV technology. But they are not, by themselves, going to make the changes required in society—there simply aren't enough of them, either. It's a nice solution for a very small minority of folks, but we can't take all of our buildings off-grid—nor should we want to. Many years ago I realized that, if PV (and other renewables) are going to make any significant contribution to society, then the utilities must become active partners in the process.

In 1984, I helped plan and execute the world's first PV-powered neighborhood in Gardner, Massachusetts. Working under contract with New England Electric, we found a subdivision with good solar exposure: thirty existing, modest, working-class ranch houses, classic post-World

War Two housing stock, various sizes from fifteen hundred to twenty-two hundred square feet with varying loads. We installed utility-interactive PV systems of 2.2 kilowatts on each roof, and we ended up with a win-win-win symbiosis.

A utility-interactive PV system uses the power grid in lieu of on-site storage. The homeowner gets free, one-hundred-percent efficient, no-maintenance, energy storage without batteries. The utility gets on-site power generation with the surplus coincident with their system's peak summer air conditioning load; they give back the surplus during off-peak or baseload periods when they're hungry for consumers just to keep the generators turning over. Society wins because the renewably generated electricity displaces polluting "conventional" sources of power.

You ask the people in the world's first PV-powered neighborhood how it's been after nearly ten years. The minority, perhaps 10 percent, really got into it—counting the solar kilowatt hours every day, and so on—but the rest blissfully reported there is no noticeable change in their lives. Most significantly, no one is unhappy. Again, the ultimate solutions will be essentially invisible to the end user. Here we have a neighborhood where thirty new electric power plants have been installed in people's yards and no one has noticed any difference in their lives! Compare that to the siting process for a coal or nuclear facility and you can begin to understand what I'm getting at.

The Sacramento Municipal Utility District (SMUD) is presently leading the way for utilities worldwide in the use of renewable energy. They're a unique experiment in democracy, where the rate-payers are the shareholders, and they can thumb their noses at the state Public Utilities Commission, from which they're exempt. Earlier managers built Rancho Seco, a particularly troubled nuclear plant, right near Sacramento, but the people of Sacramento, with the support of the current managers, voted to shut it down.

SMUD's new program goes something like: "You told us to shut down the nuke, now give us your roofs so we can install solar thermal systems and photovoltaics." SMUD plans to install some 5 megawatts of distributed PV systems on rooftops in their service territory in just the next three years.

For utility-interactive PV to really work requires a revolution in utility rate structures. The key element here is net metering, which is already the rule in Texas, Oklahoma, Maine, Massachusetts, Minnesota, and Wisconsin: whatever you generate spins the meter in your favor, and whatever you consume spins it the other way. Kilowatt hours are traded back and forth, one-for-one. At the end of the month, you pay or are paid for the difference.

With such a buy-back arrangement established, you can address each site's energy resources. Photovoltaics are certainly the most elegant and universally applicable, but wind and hydro are much more cost-effective where the resource exists.

Denis Hayes had the idea that if we were trying to do for transportation what we need to do for electricity, the government would set, say, a billion dollars on the table as a major carrot for the development of the next-generation automobile.

Denis has concluded it is impossible to get American society to willingly give up cars, so we must redefine them: four passengers to get at least sixty or eighty miles per gallon, recyclable components, automatic braking system, airbags, and the whole crash package, etc.—establish what the car of the future should be and then challenge the U.S. car companies to build it with a big reward to the one that gets there first.

Such bold new vision is what is required to bring about the watershed changes which are necessary (and inevitable) in contemporary American society—called the "paradigm shift" by PG&E's Carl Weinberg. The most significant thing is that this is also the easiest, best, and least-painful way to bring about effective change.

At present, the U.S. is just barely stumbling along. On the automotive front, Detroit fights tooth-and-nail every effort at even minor fleet efficiency improvements while car makers in other countries are already well along. It's also important to point out that these other countries have long ago adjusted to three to five dollar per gallon gasoline and are still beating us in the world markets. Enlightened consumers chose to buy more efficient foreign models and then Detroit screams for trade restraints while they lose serious amounts of money each quarter pretending it's still 1963.

My message to architects and engineers is: Look at the whole picture. In the trade press recently, there was an article hailing a custom, 9000-square-foot, architect-designed house as the latest in environmentally responsible design. Its principal claim to fame seemed to be the use of natural, nontoxic finishes on the woodwork. In the rush to commercialize "Green Architecture," no one noticed that this house consumes more energy than a small New England town.

If your goal is trying to build an environmentally responsible building, you're missing the whole point if you get all lathered up over a nonvolatile natural finish on the handrails, while you're connected to a plutonium generator power plant down the road. It's the same old "out of site, out of mind" again with a new face on it. "I'm doing all I can for the environment, my architect specified beeswax on my new woodwork—someone else will just have to figure out what we're going to do with all this radioactive waste" . . . and acid rain and oil spills and global warming and ozone depletion and unhealthy air quality and . . .

You hear a lot about sustainability these days. I've been at this since 1973, long enough to be certain that, without addressing the energy issues, you're in the weeds. All the fuss over "my milk-based paints transported in from Europe" is just a myopic distraction from the things that really matter on a global scale. True, natural-based finishes are desirable, but they fall far short of the answer. Establishing an energy infrastructure based on renewable resources is a necessary and fundamental precondition to establishing a sustainable society or to achieving sustainability at any scale. If you are not addressing the energy issues, don't even pretend that your building is environmentally responsible.

Apple CEO John Scully says the best way to predict the future is to invent it. I say, look for the "invisible solutions" that are already here, available today, and apply them. If we all just do this, the future we want will arrive before you know it.

Heating and Cooling Space

Our costliest domestic energy activity is regulating the temperature of the space we occupy. Many of us remember President Carter's cardiganed appeal from the Oval Office to turn our thermostats down in winter and up in summer. Not bad advice, but it is not the first thing we should do. Fortunately, many power companies have done their homework, and they realize that waste is the real problem. The best advice can be summed up simply: improve insulation, plug leaks, replace the glass and the heating and cooling units with the best-performing replacements. Install zoned heating and cooling with set-back thermostats so rooms are heated or cooled only when they are in use. Understand that houses built after 1910 and before 1990 assumed that energy would always be cheap.

Many are surprised to learn that the days when our electric supply runs closest to its capacity are not the coldest days of winter, but the dog days of summer, when the air conditioners are set at maximum and it is still too hot. The worst offenders, and the greatest energy consumers in this stressful time, are office buildings built—one cannot say designed, because that implies intelligence on the part of the builders—to stack workers in configurations which could scarcely be better conceived to waste cool and trap heat: solar ovens for cooking people.

The same architectural practices, applied to residential construction, produced a generation of once-fashionable flat-roofed, poorly insulated, glass-walled buildings habitable only because of their massive cooling systems. Owners of these dinosaurs are reminded of their misfortune every month when they open their heating and cooling bills. Little can be done: increase and cherish shade trees, abandon and close off the southern rooms, seek and encourage any nighttime cross-ventilation opportunities. Costlier remedies include replacing old glass with the best high-tech glass, installing or extending overhangs to block summer sun, increasing interior thermal mass in the cooler and more ventilatable areas of the house, adding a second roof and a cooling tower, and retiring the original air conditioning unit in favor of the most advanced and efficient model available. With energy trends as they are, you may expect any of these measures to pay you back in your lifetime.

The gracious houses built before architects and mass-production ruined the American notion of a suitable dwelling handle summer's heat better. In areas with hot summers, stately deciduous trees are already in place, and in fact whole neighborhoods enjoy temperatures several degrees below those noted at the airport. For these older homes, the standard measures can be applied very effectively, and will improve thermal performance in summer and winter: Beef up attic and, if possible, in-wall insulation. Make sure that infiltration of cold air in winter is controlled by weather-stripping and caulking windows. Maintain and improve window shades, especially on the south side for summer cooling and north side for winter heating. Employ the tools intended by the house's builders: maintain storm windows and sashes, cross-ventilate in summer and close off unused rooms in winter, and use door-snakes (those long, bean- or sand-filled socks that block the crack below the door.) Replace aging sash windows with the best available high-tech windows, knowing that your investment will repay you handsomely in a very few years. Install new cooling and heating units before the old ones are worn out, secure in the knowledge that here, too, your investment will be quickly repaid.

Many utilities offer free home energy audits, the most comprehensive of which employ infrared cameras that ferret out leaks and weak spots in your home's defenses against the frigid outdoors. This technique works best in winter, but improvements in wintertime performance often apply as well in summer, when we wish to invert the house's thermal behavior and keep heat outside. The results may help you decide which measures will be most cost-effective. For example, if heat transfer is happening primarily due to single pane windows, all the caulking and weather-stripping in the world will make a very small impact.

An American Fetish: Hoarding Hot Water

Installing a solar hot water system is undeniably one of the best resource- and money-saving actions you can take. Choose one that is appropriate for your climate, and it will pay for itself in five years or less.

Let me ask some personal questions. Are you a hot water hoarder? Do you have, somewhere in your abode, a tank full of hot water patiently awaiting your call? Most Americans do; most everyone in the rest of the world does not. It's one of our strangest cultural quirks. How long does it take from the time you turn on the hot water tap until water at its hottest comes out?

There are three measures we may apply to this state of affairs. The first, as already pointed out, is to install solar hot water heating. By using solar energy to heat water, we replace a highly significant energy load with a free source, and we are put in touch in a very important way with the day-to-day state of our planet. When we need hot water, we must devote

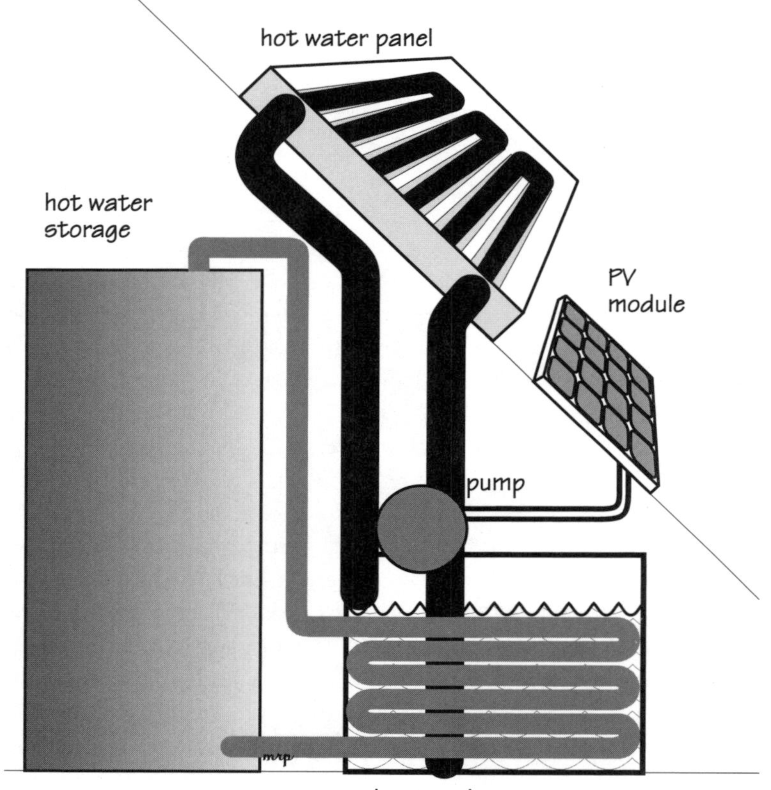

Figure 6. Solar hot water heater. Sun shining on the panel heats the fluid. At the same time, sun shining on the module runs the pump, which circulates the heated fluid. When the fluid in the heat exchanger gets hot enough, it starts to heat water in the hot water storage tank by thermosiphon: hot water rises and is replaced by cold water. When the sun goes down, the pump stops, and the fluid drains out of the panel, so there is no danger of freezing.

just an instant's thought to our recent sun and water history. Has it been sunny? If so, there will be an abundance of hot water, but if not, it would be wiser to conserve. Has there been much hot water use since the last sunny period? If so, we would be wise to use sparingly or defer our water use, but if not, we may exercise our normal, water-conscious practices. An occasional cold shower will help us refine our estimating skills. I like to measure results, so I added a couple of thermometers to the system; now I can watch the water temperature rise on a sunny day, monitor hot water availability, and gloat over my energy savings.

Many solar hot water systems have gas- or electric-assisted back-up for cloudy periods and families that like to take showers when they wake up. Although these systems still hoard the hot water, in sunny times their electricity consumption is dramatically reduced. By putting a timer on the back-up (see the third hot water cost-cutting measure below) we may avoid electrically heating water that would be heated by the sun a little later. For example, as we take a morning shower, cold water replaces the hot water we use in the bottom of the hot water heater; an electric system will sense its presence and start expensively heating it even though we may not need more hot water until evening, by which time the sun would have heated it for free.

The second measure, if we are offended by hoarding hot water or are unable for technical reasons to install solar hot water, is to disconnect the energy-guzzling hot water heater and install local hot water sources like the ones used everywhere else in the world where hot water is served. These devices, called demand or instantaneous water heaters, heat water only when it flows through them. By placing an instantaneous water heater close to the delivery point, within a few feet of the sink or shower, water wasted while waiting for the hot stuff is minimized. Even if the new unit is installed at a distance from the kitchen sink, shower, and bathroom sink, for instance, in the closet or basement where the old hoarder used to be and where plumbing already exists, hot water energy expenses will decrease because keeping a standby quantity of hot water is costly.

The third measure, and simplest of all, is to turn down the thermostat and install a timer on the hot water heater, so that it brings water no higher than the temperature at which we will use it, and only during times we habitually require it. Hot water heaters are generally set too hot, so we must temper their output with cold water; this makes no sense, and costs money. Most conventional hot water heaters recover in an hour or less, so allowing an hour's lead time is adequate. My system turns itself on once a day, late in the afternoon—at the end of the solar day so that, if there has been sun, the water heater's thermostat keeps it from using energy. Morning ablutions are more expensive because the solar-heated supply is at its lowest ebb then, and grid energy will almost be needed for hot water.

If you have not already done this, please do it right now: Turn your hot water thermostat so it is as hot as you ever want tap water, but no hotter. One hundred twenty degrees Fahrenheit is a good maximum. One hundred thirty degrees Fahrenheit, only ten degrees hotter (and hot enough to scald) costs about 25 percent more.

The Outrageous Cost of Refrigeration

Why does a home refrigerator cost so much to operate? The short answer is, buy a super-efficient refrigerator and it costs much less; buy the more expensive, more efficient unit and seven or eight years from now, comparing equipment cost (the original purchase price) plus the operating cost (price paid for the electricity consumed), you will break even and be a resource hero. If you keep the refrigerator for twenty years, your total cost will be half that of the inefficient unit.

In the late 1960s the lifetime cost of a new refrigerator was expected to be about as much as the unit itself cost, and therefore energy efficiency was not an important issue. Today that same refrigerator, humming happily away in your kitchen, can be expected to consume energy costing six to eight times its original price before it fails. Curiously, refrigerator efficiency decreased spectacularly during the decades following their introduction to the consumer marketplace, becoming the household's most feature-laden showpiece appliance—and an energy and ecological horror. Despite their primary mission of keeping cold, they use copious energy to heat themselves to keep coils and doors frost-free. The CFCs in the refrigerant and embodied in the insulation endanger ozone and therefore decrease our ability to enjoy the sun; the waste heat dumped into the house by poorly designed refrigerators further burdens our domestic cooling systems, consuming yet more energy. There is only one remedy for such appliances: replace them.

Ask a dozen engineers why refrigerators are built with their compressors and heat exchangers underneath their cold chambers, and get a dozen answers, but not one of them will say, "because it works better that way." Because it doesn't. Maybe it moves the unit's weight lower, thus making it more stable so it is less liable to fall over on people. Maybe it makes the unit easier to ship. Maybe it looks more pleasing to the householder's eye. Maybe it made the production line easier to design, and saved the manufacturer 25 cents per unit. But it makes the refrigerator inefficient.

Conventional refrigerators work by compressing a normally gaseous compound until it becomes liquid, a process that uses electricity and gives off heat. The liquid refrigerant is kept compressed while it dumps the heat of compression into the room, then is allowed to expand in coils inside the refrigerated chamber. Expansion requires heat, and so the coils in which the refrigerant expands draw heat out of the chamber, cooling that enclosed space. The vapor circulates back to the compressor, where it is compressed again, using electricity and giving off heat, and around and around it goes. Remember that heat rises, so if all this compressing and radiating takes place under the cold part, the heat will rise and interfere with the cooling, making the whole thing less efficient, particularly if the cold part is imperfectly insulated, as it often is. Add a few more tricks, like a defrost system that periodically heats things up in the cold part, heat tape to keep the door seals free of frost, and other wrinkles that sacrifice energy for convenience and easy production, and you have an energy hog that ranks just below the hot water hoarder for energy wastefulness.

Michael Reynolds is developing an unconventional thermal mass refrigerator. The thermal is located in the chamber's walls, in the freezer box shelf, and in the refrigerant stored in cases at the bottom where the coolness pools. In cold months, when outside temperature is at or below the target temperature for the freezer, the operator can open the box's skylight. In warmer months, a small refrigeration unit kicks in to keep the temperature low. This design uses a static refrigerant in the form of beer or soft drinks instead of the toxic used in active refrigerators. So far the primary development problem has been refrigerant loss to thirsty lab colleagues.

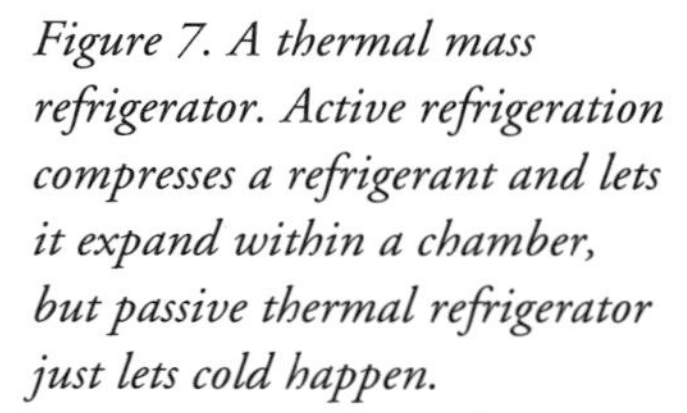
Figure 7. A thermal mass refrigerator. Active refrigeration compresses a refrigerant and lets it expand within a chamber, but passive thermal refrigerator just lets cold happen.

It has long been known that refrigerators are inefficient. For the present generation of conventional refrigerators, designers have done away with some of the more wasteful tricks and improved the seals and insulation. Many large utilities have run campaigns to get the really inefficient refrigerators, those made before 1980, out of service by offering to get rid of them in favor of newer models, even subsidizing the purchase. Yet we know that even these new models could be improved dramatically, because Sun Frost builds units that are six times as efficient using the same technology, units that (guess what!) have their compressors above the cold part; the Sun Frosts are also extremely well insulated.

Now here's a joke. The electric industry has offered a golden carrot: one million dollars to the company that can build a refrigerator as efficient as the Sun Frost—but Sun Frost cannot compete because they did not make enough refrigerators in the qualifying year. The reason for that, besides the fact that the Sun Frost gang build their refrigerators one at a time in an old cannery building, is that it costs quite a bit more to build a really efficient refrigerator, and buyers are afraid to pay so much as an initial outlay.

The equation is really quite simple. If two refrigerators, a Sun Frost (SF) and a best-of-breed conventional unit (BOB), are tested side by side, we will want answers to four questions: (1) How much do they cost to buy? (2) How much do they cost to run? (3) Do they work equally well? (4) Do they last equally long?

Suppose we establish that they work equally well and last equally long. (That's not really true; the Sun Frost works quite a bit better and is easier to service, so it might last indefinitely. But let's suppose equality just to make the other comparison easier.) The answer to our first question is, unit SF costs five times as much to buy as unit BOB. This staggering difference stops most buyers from considering further; sticker shock dulls their wits, and another BOB unit is sold. But wait! Unit BOB costs six times as much to operate at present energy costs! Said another way, if we compare the whole package cost (refrigerator price plus electricity used to date) the SF unit's total falls below that of the BOB package after eight years in service. If we assume that energy costs keep going up at about 10

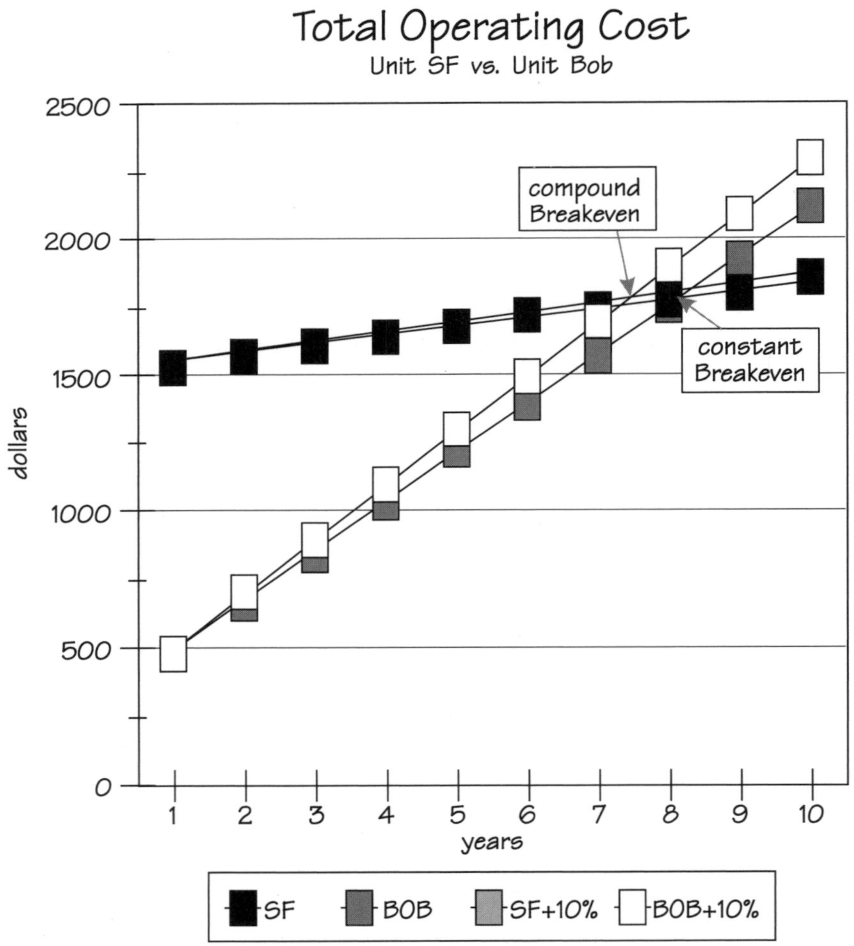

Figure 8. If electricity costs keep going up at their present rate, the superefficient Sun Frost unit starts costing less after seven years.

percent per year, compounded annually—which accounts for recent increases fairly accurately—the payback period, the point at which the whole SF package cost is less than BOB's, is a little more than seven years.

We bumped up against this issue in our earlier discussion of lighting, and we will again. Pay now, or pay later; by paying more now, we pay less in the long term and have the satisfaction of knowing that we have made a positive decision for the planet. Do consumers pick up the big ticket? Not enough of us—at least, not yet.

Proudly standing alone: Catherine, Eva, and Lilia's story

Catherine Downey lives with her two daughters, Eva and Lilia Dubey, in their owner-built home above a tributary of Opaeka'a Stream on Hawaii's Garden Island of Kauai, five miles east of the rainiest spot in the world, Mount Waiale'ale. Just seven months before our talk, Kauai was battered by Iniki, its second killer hurricane in eleven years.

Catherine: I changed the brake shoes on my car once, and decided I didn't want to be mechanical. As a naturopathic doctor, I'm a scientist only by default. As a midwife and naturopath, I'm interested in natural living.

We try to live naturally. The bugs bug us, especially the big flying cockroaches we call B-52s. Roach hotels work—Tora! Tora! bes' kin', bra!—and they get their feet stuck. I won't use poisons, except for boric acid traps for the ants. Ants are really bad when it rains. We're trying to get away from using plastic bottles because I just read that Americans are throwing away two and a half million a day.

I believe in midwifery, in its spiritual basis. When white man and red man met, the patriarchy was supposed to meet the matriarchy and create a more intuitive culture, but somehow we blew it . . . and that's why we still worship a god up in the sky. In my work, I bring together modern things, IVs and medications, and ancient midwife traditions. I have a doppler that I use to hear the baby's heart; I imagine that ancient midwifes would have loved to have that, to be able to say to the mother, "Roll over on the other side; you're pinching the cord."

For me, this consciousness started when I became a vegetarian when I was twenty-one. (I'm not strict; I'll eat fish sometimes, and a little chicken . . .)

Eva : You don't eat much chicken. But Lili eats hamburgers!

Catherine : Yes, she does.

To the best of my ability and knowledge, I built this house as well as I could. I moved very little earth, and the lumber came from farmed

Catherine and Eva check the battery box.

trees. The house is built of untreated lumber, cement, and steel. If I did it over, I would use more cement. It was important to build on marginal land, and let the good land alone for farming. We have the technology to build on hills, and we should.

I left the trees—Java plum, Formosan koa, volunteer bananas—by the stream. There are probably few native species, but the birds love the plums. My neighbors cleared all their trees. They say to me, "When you going to clear your land, make it useful?" I think I will leave it. Now, I get all the birds.

Except for the koa, which broke in half, Iniki (the hurricane) left nothing but sticks. They looked like they were dead. "You just as well bulldoze it now, it's dead, and take to the dump," the neighbors said, but look at it now, in less than a year, it's grown back.

During the storm, we stayed downstairs, stood on the porch and watched. It was quieter there. At first the storm came from the east and broke a few windows upstairs. I think the fact that there's so much screen, and the wind could blow in and out again, saved the house. It sounded like a big train coming, and the trees in the distance were bending over, going *Eeyah!* I could tell from that palm which way the wind was blowing. It shifted more south, and then at the end came from the west. Pieces of neighbor's roofs were flying over and settling into our valley where the wind was not so strong. I remember seeing birds blown in by the gusts and taking refuge down by the stream; I remember Eva saying, "Mom, there are still birds!" Right across the stream, that house moved fifteen feet off its foundation with the people still in it. Whew! We just had some water damage, and lots of leaves in the house.

After Iniki they wouldn't let us scavenge at the dump. I needed a second toilet, and there were people whose houses had been blown apart—they'd thrown everything into dumpsters, and there were lots of good toilets, but I had to buy one new. We just started recycling on Kauai. Before, they said there was no way to do it. I say, you live on an island, you better recycle! The island generated five years' worth of garbage in that one day. They could have called it the Garbage Island.

I've lived on Kauai for thirteen years now, and both my daughters were born here. When I bought this land, close to where my ex-husband lives, both he and my boyfriend were upset, but I did it for the girls. This way, their dad is close; he's got the piano, so Eva can go up there to practice. They each have one bike, one boogie board, and they get to have their dad. It works out well.

We lived in a yurt on the land while we built this house—I loved it! The electric company charged three thousand dollars to install our electricity, and they made me give easements to my two neighbors before they would do it. I was pissed, because I wanted to spend that money on solar. When the electricity went down after Iniki, the insurance company had a loss-of-use clause that let people buy generators so they could stay living in their houses, and there were all these noisy gas generators around. I told them I wanted to buy a solar generator, and they said okay, just show us the receipts. That's how I got my system.

I built this house as the owner-builder, but I had a lot of help; I didn't pound very many nails. We had the usual inspections, foundation, framing. In Hawaii, you have to have a licensed plumber and electrician do those parts. We never mentioned to the inspector the house was on solar, but we didn't hide it either, and of course he could see the modules, and he didn't say a word. The AC smoke detector required by the building code didn't like the electricity out of the inverter, so I disconnected it and put in one with a battery after he left.

Now, I use the grid as my back-up, and we run everything on electricity from the four modules on the roof. I watch the meter, and when it says my batteries are 90 percent full, and I know it's going to be sunny, I do a wash. Sometimes, but seldom, when somebody turns on something like the vacuum in the morning when the batteries are below 85 percent, I can see by the meter lights that we're using the grid. I don't know how much grid power we use, but I can tell you it isn't much.

Eva : Because Mom's nagging me all the time to turn off the lights!

Catherine : We're adjusting to living within the amount we produce. I don't know too much about how these things work, so Ross, the installer, wrote everything inside the box, and gave me lessons.

What got me interested in solar power? I thought it would be so neat to refrigerate with the sun. I finally got my Sun Frost, and we had to remodel the kitchen because it's so big. I want to add at least eight more modules and four more batteries; if I add more batteries, I'll have to build another battery box. Next, I'm interested in solar water heating. I have an instantaneous water heater, but I'm paying too much for gas. And an electric car . . . I'm committed to being oil-free by the year 2000.

Hawaii is one of the few states that licenses naturopathic doctors. There are two others on the island. My specialty is women and children, and I'm the only one that does home births. I've done, maybe, four hundred, and the one this morning was so beautiful! But the more you do, the more you see problems, and appreciate that it's life and death. Here's what I did so far today: I was at that birth all night, and the baby was born at 4:33 this morning. At nine I was in a canoe race; we came in third overall and won our division—the masters, women over thirty-five. I guess we're masters because we've mastered the art of being athletes. We're not as strong as the twenty-year-olds, but we're better paddlers. Look at this silver medal from the 1991 statewide Hawaiian Canoe Racing Association competition . . . Now I'm paddling with the Kaiola club down in Nawili'wili. We have a smart coach, and this year we'll win!

Wise Use and Phantom Loads

Phantom loads are small but constant energy drains—clocks, transformers, and other miscellaneous energy guzzlers. Felicia Cowden calls these phantom loads the energy criminals of the small appliance world. (You will find Felicia and Charlie Cowden's story in chap. 4.) If you have office or entertainment equipment plugged directly into your wall plugs, you waste quite a lot of power even when the equipment is not in use. Instant-on televisions and other gear with remote controls are always on, even when they appear to be powered off, because they are always eagerly poised for the call of the remote control unit. The little plug-boxes that power calculators, clocks, rechargers, and all manner of other small appliances are also horribly wasteful. The first thing to do for these is put them on plug strips or switched outlets so that they may be completely turned off when not in use. (This strategy is not particularly appropriate for clocks.)

Small electronic devices that require transformers are themselves often quite efficient, especially because they use low-voltage current. The real thief is the little black box, the transformer: this cheaply made inductive monster wastes about 80 percent of the energy it consumes even when the device it feeds is turned off. Its job is to step the house current voltage down (transforming it to a lower voltage) and then to convert or rectify it from Alternating Current (AC) to Direct Current (DC) if required by the device it powers. The wasted energy turns into a surprising amount of heat and electromagnetic radiation (EMR). Existing electronic devices can easily be converted to higher efficiency by connecting them directly to 12-volt DC circuits if (and this is an important if) they use 12 volts DC, which many of them do, and you have 12-volt electricity available. The name-plate on the transformer will specify output voltage; anything in the range of 9 to 15 volts DC will probably work fine. If in doubt—especially if the device is costly—check the manual or call the technicians.

Phantom loads are certainly plug-load non grata in any energy self-sufficient house, but they should be considered just as offensive on-the-grid, since the only difference is that the offense occurs in somebody else's backyard. Beyond the big three of potential energy savings discussed above (space conditioning, hot water heating, and refrigeration) eliminating phantom loads, reducing plug-loads, and buying appliances that are designed for efficiency can save the most domestic energy. Here is a great test for on-the-gridders: Some comfortable day, turn off or unplug everything you can think of, including the hot water heater and refrigerator, then

Off-the-Grid Day

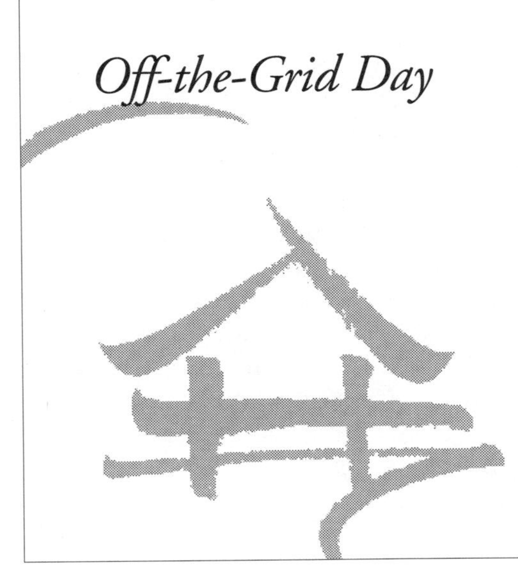

A great day to conduct the phantom load test is Off-the-Grid Day, held annually since 1991 on a Sunday in mid-October. On this day, you might try an even more radical test: after taking whatever precautions you know are required after a power failure, find and turn off your home's master electrical switch. It is a good idea to know how to do this anyway. Then go on about your business, carefully noting any missing conveniences, and the lengths to which you and your family must go to provide alternatives. Be aware, for example, of managing refrigerator access. You cannot afford to open it and contemplate its contents, so you must decide what you want before the door opens.

This off-the-grid experience will help refine your perceptions of your electrical dependency, and (if you are joined by enough friends) will send a major message to the utilities about your energy awareness and willingness to conserve.

read your electric meter. (Your electric company should be happy to teach you how to read your meter if you do not know how.) Read it again an hour later. The difference is the electricity stolen from you every hour of every day by your household energy criminals.

The road to town: Noel Perrin's story

No discussion of energy dependence can neglect our love for the motor vehicle. Energy expenditures on internal combustion engines are huge, and such engines are still one of the dirtiest power sources ever devised, despite our recent (and halfhearted) efforts to clean them up. This is an issue of such magnitude that many people surrender to it completely and make no effort to cope. A few brave souls, among them Noel Perrin, have tackled the issue head on.

Noel Perrin lives in a Federalist farmhouse near Thetford Center, Vermont. He teaches environmental literature at Dartmouth, and his subject matter has led him to think deeply about energy questions, and suit his actions to his conclusions. His experience with a solar electric vehicle has prompted him to take the next step: building an off-the-grid home.

Noel Perrin and his solar electric vehicle (photo by Robert Pope, used with permission).

We are a restless folk. A rich old woman of my acquaintance wants to be taken for a daily drive in her car—often she's out for two hours. She has a perfect right to like motion. But wouldn't it be better, using the same amount of money, and without environmental damage, to get a nice matched pair of greys, a comfortable rig, and perhaps an Amish driver-keeper to maintain horses and rig, and take her for her daily drive?

We need to get real. It's as John Jerome says in *The Death of the Automobile*: Everybody knew the automatic shift wasted power, so we just had to add more power. Now we must learn that power is not inexhaustible, and is unacceptably costly.

I came to live as I do because I love this old house, and because of an accidental character trait—being frugal. I came in 1959 with no prescience or thought of the environment, to teach English Literature. Living here, and heating with wood, I became aware of a productive closed loop, wherein I was clearing my woods, and saving money, and getting firewood. By Earth Day in 1970, I was selling excess firewood off the back of my pickup in Hanover to collegians interested in practicing seduction by firelight.

The wood harvesting could be cleaner, and will be when a friend develops his portable electric chain saw system. An unfortunate fact of internal combustion engines is that the smaller they are, the dirtier. We bring the wood in, and perform other farm tasks, with a diesel farm tractor. We may lie slightly more lightly on the earth than a typical middle class family, but we are still harder on the environment than two or three families in India.

That first Earth Day was another major force in my life. Regular classes at Dartmouth were cancelled in favor of a special curriculum, and I attended a class, "Pollution in Dartmouth's Backyard," which revealed that we were burying low-level radioactive waste on college-owned land. To this day, there is a striking discrepancy between the preachings of the Environmental Studies faculty, and the practices of the college (and I daresay this would be true of any college with such a faculty). Why not follow the advice of the environmentalists on campus? And commence behaving as we know the First World must? Or, if we are not believed, why are there no firings or gaggings? In fact, what we say *is* believed, but only as dry facts. Even the best education moves us only a little; events, like Three Mile Island, move us a lot.

Does that mean we must have more and better disasters before we commit to a more rational approach to energy? I hope that an abstract understanding of extinction will prove enough. If I'm wrong, I'd rather not know it. My wife asked her fourth- through eighth-graders at Lyndonville to write an essay about "the Earth our Garden" and they all wrote about toxic waste dumps. To be totally environmentally aware as a child would be too awful, as was the idea of "mutually assured destructive capability" for children in the last generations.

One must present the information without preaching; most of us prefer not to be preached to. Since television is the source of half the world's information, Population Communications International has been financing sitcoms and soap operas seeded with population and energy awareness messages, which have been produced by O Globo in Brazil and syndicated internationally. I think the best operating principle is, whatever works, works.

Most of us would rather do something (like burn down a billboard, and there are plenty of small wooden billboards for those who wish to amuse themselves that way) rather than meeting to legislate against billboards. I wouldn't claim that my work, exploring Edward Abbey and his peers with my college seniors, is more powerful than teaching environmental science, but it should be an important way to spread awareness. A student brought me a letter from his father in response to my assignment to conduct a family environmental assessment: "Son, may I remind you how we came to own our fourth car . . ." We need not look too far, yet, for things to improve upon. Nor, I think, need we look too far ahead; it works best for me to look forward no more than two or three years.

A few years ago I was taken to task by my students for commuting thirteen miles from home to work, and I decided to correct my error. Pollution seems the most threatening risk, so I decided on an Electric Vehicle (EV) recharged photovoltaically by a grid-backed system. I wrote about finding and retrieving the EV in *Solo*, but the story about the rest of the system may be more interesting here. I wanted enough pollution-free electricity to recharge my car, and I preferred not to buy twenty or thirty extra batteries besides those in the car. Since I wanted to connect my photovoltaic array to the power lines, this caused some initial difficulty for the utility, which they were quick to share. They wanted me to carry five million dollars worth of liability insurance, which was more than I wanted to carry, and they required absolute assurance that my little system would shut down in case of grid failure, so that line men would not be at risk. (I wanted that assurance too, because I know some of the linemen and women.) The utility executives were at last convinced that a standard homeowner's insurance policy would suffice, and that my inverter's fail-safe mechanism would offer protection. As it has turned out, it has been grid power that has caused the problems: twice, line surges have blown up my inverter.

I have two meters; theirs records the electricity I consume, which I pay for at the standard rate, and mine shows my generation, for which I am paid a lower rate (which is fair; I don't maintain a distribution system). On a winter day, I may generate between 800 and 900 watts in the four peak sun hours between ten and two. That comes to about 3 kilowatt-hours a day. My EV has a range of forty-five to sixty miles on level ground, and so in summer, in theory, I might be self-sufficient for transportation. You may have noticed that around here it's mostly not level, and mostly not summer.

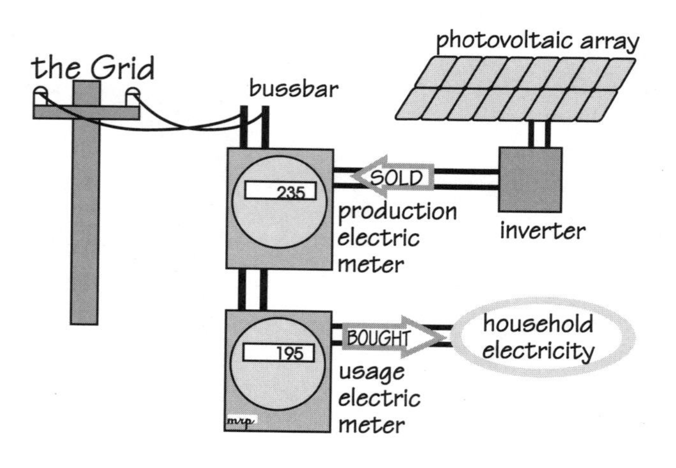

Figure 9. Utility intertie systems often have two electric meters. The utility pays you for what you produce, and you pay it for what you use.

It can get very cold here, so even this winter the pipes froze. We can close off all of this house but the stove room, where we and the cat stay comfortable, close to the woodstove.

We are having a new home built (we've done all the nailing) which will be more sensible: a sixteen-by-twenty cabin with a sleeping loft and a writing room at each end, oriented due south, of course. It will be off-the-grid, and powered by eight panels and twelve batteries, and as well-insulated and glazed as technology permits. It will be hard to leave this lovely old house.

A Cleaner Obsession

Electricity is a clean and well-behaved form of energy that we can harvest ourselves with almost no effort. The forces underlying many forms of energy are uncertain and obscure, but electricity is elemental in its simplicity. By mastering its measurement and management, we become entranced by electricity's proximity to some of the most basic properties of matter and life.

In the building trades, domestic electricity is a relative newcomer, and as a result its practices and procedures are unhampered by the kinds of historical conventions and anachronisms that plague plumbing. Electricity is not sneaky, the way water is. Likewise, the materials you need—wire, boxes, outlets, switches, and meters—are straightforward and comprehensible, and the tools—hammer, drill, stripper, nipper, knife—are easily managed. Working with electricity does not require strength, but does require intelligence. Even all-thumbed people can put electricity to work for them. It is an extra satisfaction that our efforts in improving our homestead electrical systems also remove a weight from the overburdened planet.

CHAPTER 3

HARVESTING OUR OWN POWER

GENERATING SOME OF MY OWN POWER IS ONE OF MY MOST ENTERTAINING AND rewarding hobbies. Everyone who uses a light-powered calculator, a solar flashlight, or a stand-alone walkway lamp is an alternative energy hobbyist. What an engaging concept: something as essential as energy comes to us for nothing more than basking in the sun! Photovoltaically powered call boxes are becoming a familiar and reassuring presence along the nation's highways. In addition to their life-saving mission, they are prominent representatives of independent energy, showing that renewable energy is where you need it, and it works.

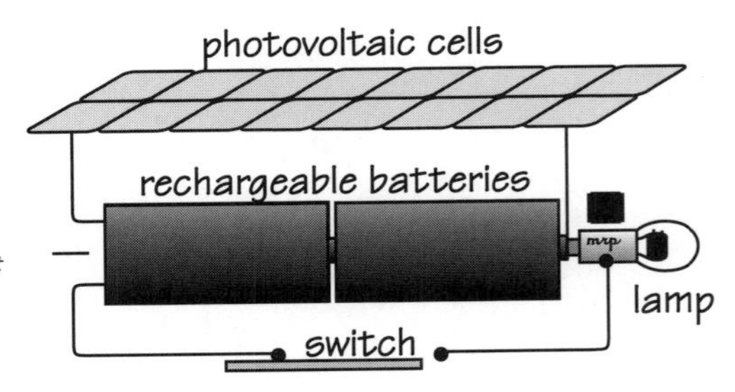

Figure 10. A solar flashlight, and all sun-powered systems share four elements: photovoltaic cells (1) harvest solar energy and store it in rechargeable batteries (2), where it waits until the switch (3) closes the circuit and sends energy to the lamp (4).

The commonest sustainable energy sources—energy sources not used up over time—are photovoltaic (PV), micro-hydro, and wind, all of which produce electricity; and biomass, which produces either electricity or methane.

Many schemes for harnessing other potential sustainable energy sources—geothermal, ocean waves, the temperature difference between the ocean's surface and its depths (called the thermocline)—are being explored. So far, these sources have been found to be impractical because of their sensitive, remote, or unusual locations, because of difficulties in making them large enough to be economical, or because the sources are intermittent or difficult to work with. Equipment at pilot geothermal plants, for example, is failing much earlier than predicted due to corrosion.

Harvesting Sunshine: Photovoltaic cells

In the 1950s, when the photoelectric effect was explored, it was an expensive laboratory curiosity, yet its promise, an unlimited supply of electricity, was very important. The world's finest technology was applied to put this scientific breakthrough to use, with a clear objective: develop a reliable energy source which (1) liberates as many electrons as possible (2) with the cheapest and most abundant materials (3) manufactured in a nonpolluting way. The problem is not trivial because semiconductors are noteworthy primarily for their smallness, whereas PVs harvest electricity from sunlight in direct proportion to their size. Materials scientists have dramatically decreased the costs and increased efficiency at every step, growing purer and larger silicon boules in pressurized ovens, slicing them with ever-thinner diamond saws so less material is lost to the saw kerf, doping them, and assembling them automatically into finished cells. Amorphous and multicrystalline cells can now be grown in thin sheets, eliminating the boules and wasteful slicing; in the future this process may allow further decreases in cost. Other materials besides silicon, and many doping and fabrication strategies, have been tried, but there appear to be no profound breakthroughs ahead. Costs will continue to decrease because of the technological learning curve, as materials and techniques are fine-tuned, and modules are produced in ever-larger volumes.

A good way to measure the cost-effectiveness of a solar cell is in dollars per peak watt: the cost divided by the cell's best energy output. The first PV cells were astonishingly expensive, costing more than one thousand dollars per peak watt in 1958. The first customer was, of course, the space program. Since then, cost has decreased by roughly an order of magnitude (a multiple often) every fifteen years. As with many new technologies, the price has dropped so dramatically in such a short period of time that it is impossible to foretell what may happen in the future. By

The Photovoltaic Effect

We are all made of light, but we do not really understand what light is. The best we can do is to make models, and science has made two: light as a *wave*, or light as a *particle*. A single wave of light has a certain vibrational frequency. Light of many frequencies makes up the *electromagnetic spectrum*, of which visible light is a narrow band where we perceive frequency as color: red light, for instance, has a higher frequency than violet light. Below violet we find lower frequencies, including ultraviolet and x-ray. On the higher end of the electromagnetic spectrum, above the visible bands, we find higher frequencies: infrared, radar, microwaves, and radio. This wave model works well when we are trying to understand how light travels.

But not all of light's behavior can be explained using the wave model, and so we use the notion of the photon: a small particle of energy which manifests its existence as light. Higher energy photons are associated with higher frequency waves, so blue light is more energetic than red light. We know that light can be emitted when materials have energy applied to them (for example, by heating a tungsten filament in an old-style lightbulb), and our photon model explains this: when an electron associated with one of the atoms in the filament is whacked hard enough by the electrons flowing through, it emits a packet of energy called a *photon*.

Science often proceeds by formulating an understanding of a phenomenon, a theory, then seeing if the theory can be turned upside down: if energy can generate light, can light generate energy? This is precisely how the photovoltaic (PV) effect was found. Whack an atom with a sufficiently energetic photon, and one of its outer *electrons* may go wandering off by itself. If in its random walk it escapes to a conductor, the moving electron (together with many, many of its peers) may be detected and employed as electricity. The lifespan of a wandering electron is usually quite short, but some substances are more hospitable to free electrons than others. *Electrons* and *holes*, the *carriers* of moving electricity, are massless energy packets, and do not behave like particles. A hole is the positively charged theoretical opposite of the electron, explained by perfectly sober physicists as a place that "wants" an electron. The whereabouts of carriers can never be accurately predicted because of their masslessness; rather, we may describe their probable behavior. Carriers are seldom measurable singly, and so probability provides a good tool for describing their group behavior.

Conducting and Resisting

All materials conduct electricity: some very nicely (gold, copper), some well enough (aluminum, carbon, tungsten), some poorly (seawater, skin, wood), and some only under duress (air, rubber, plastic, mica, glass). These latter substances conduct so poorly that they are called *insulators*. *Conductance* measures a material's ability to allow electricity to flow through it: gold has very high conductance. *Resistance* is the opposite of conductance, and so we say that mica's resistance is quite high. Generally, a material is *conductive* if its atoms have only a few electrons in the outer reactive shell or energy level, called the *valence band*. For most common elements, those in the center rows of the periodic table of elements, eight electrons fill the valence band. If an element's valence band is full or nearly full, it conducts poorly. The inert gases—helium, neon, and argon, for example—have full valence bands and are unreactive and nonconducting, which is why they are called inert. Metals have one, two, or three valence electrons, which they hold loosely, and metals are therefore excellent conductors. Elements with incomplete valence bands tend to combine with each other by sharing valence electrons in ways which fill the valence band, and the completeness of their arrangement determines the stability of the compound. Silicon has four valence electrons, and oxygen

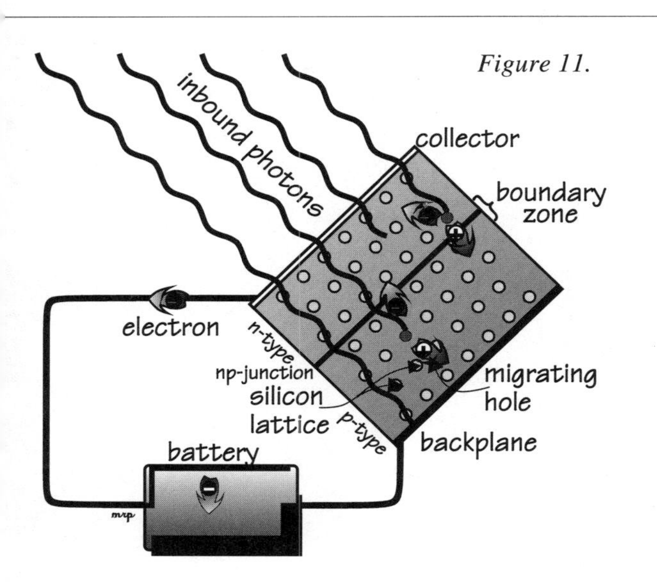

Figure 11.

has two, so silicon dioxide (SiO_2, the primary component of sand) is a stable compound because in each molecule two oxygens and a silicon share their eight electrons quite companionably. Silicon, like the other elements in the fourth column of the periodic table, is a *semiconductor*, because each atom bonds tetrahedrally with four neighbors and shares its four valence electrons with them. Under the right conditions, these semiconductors change resistance dramatically, switching in an instant from insulators to conductors, because so many nuclei share so many valence electrons.

From Light to Energy

In the 1950s scientists tinkering with semiconductors found that by introducing small, minutely controlled amounts of certain impurities called *dopants* to the semiconductor matrix, the density of free electrons could be shepherded and controlled. The dopants, similar enough in structure and valence to fit into the matrix, have one electron more or less than the semiconductor; for example, the process of doping silicon, which has four valence electrons, with phosphorus, which has five valence electrons, produces a (*negative*) *n-type* semiconductor, with an extra electron which can be dislodged easily. Aluminum, boron, indium, and gallium have only three valence electrons, and so a semiconductor doped with them is (*positive*) *p-type*, and has a hole where each missing electron wants to be. These holes behave just like electrons, except that they have an opposite, positive charge. Although loosely bonded or extra carriers—either electrons or holes—exist in a substance, the matrix is still neutral electrically, because every carrier is matched by a proton in the nucleus.

The magic begins when the two semiconductor types are intimately joined in a pn-junction, and the carriers provided by the dopants are free to wander. Being of opposite charge, they move toward each other, and may cross the junction, depleting the region they came from and transferring their charge to their new region. This produces an electric field, called a *gradient*, which quickly reaches equilibrium with the force of attraction between excess carriers. This gradient becomes a permanent part of the semiconductor device, a kind of slope that carriers tend to slide across when they get close to the junction.

Light striking a photovoltaic cell is, at the particle level, a bombardment of photons striking atoms, which give up electrons. When an electron gets lopped off an atom, it leaves behind a hole, which has an equal and opposite charge. Both carriers, the electron with its negative charge and the hole with its positive charge, begin a random walk generally away from each other because the gradient slopes "downward" in opposite directions for each. If either carrier wanders across the pn-junction where the two types of semiconductor meet, the gradient and the nature of the semiconductor material discourage it from recrossing. A portion of carriers that cross this junction can be harvested by completing a circuit from the collecting grid on the cell's surface to its metal backplane. In the cell, the light pumps electrons out one side of the cell, through the circuit, and back to the other side, energizing any electrical devices (like the battery in the diagram) found along the way.

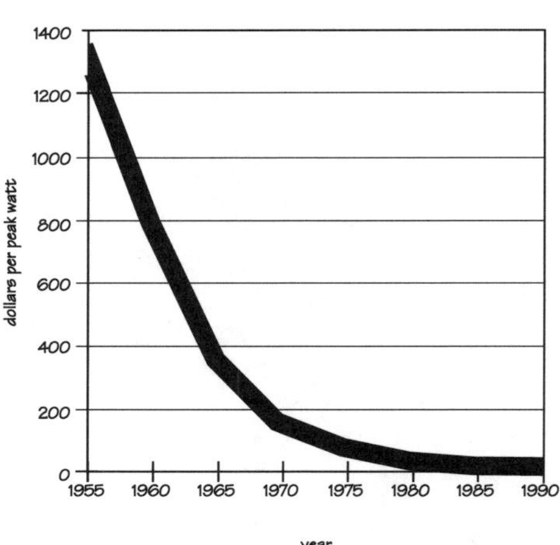

Figure 12. Photovoltaic prices have decreased dramatically since 1955, but appear to be levelling out between one and three dollars per peak watt.

extrapolating this trend, I would like to imagine that photovoltaic modules will sell for one dollar per peak watt by the year 2000, but it is reasonable to assume that at some point the gains will start to flatten out. A home photovoltaic system includes more than the PV cells, and so a system driven by four-dollar per peak-watt cells would produce a kilowatt of electricity for about thirty-two cents.

Solar Politics

The obstacles to developing photovoltaic power as a competitive energy source are formidable, and unfortunately include political and military, as well as technical, complications. The Department of Energy, which supervises and funds research into new energy sources, has devoted a small portion of its research funds to renewable sources, preferring to investigate potentially strategic energy sources and to support glamorous Big $cience like nuclear and plasma research—projects with possible military significance. The National Renewable Energy Laboratories (NREL) research program, sponsored by the government, has focussed its solar cell research on five out of hundreds of possible semiconductor strategies: amorphous silicon, Copper-Indium-Selenium, Indium Phosphide, Gallium arsenide, and cadmium telluride. The reasons for choosing only these may not be purely scientific; in fact, there are good reasons why none of this research should be carried forward. Amorphous silicon, in which the tetrahedral lattice is chaotic—and all amorphous semiconductors—have problems with age and large area uniformity, which means it is hard to make them big, cheap, and tough enough to compete with crystalline silicon. Apart from technical problems, Copper-Indium-Selenium (CIS) cells consist of a strategic metal, a rare metal, and a poison; in fact, indium is so rare that world supplies, if entirely exhausted in order to manufacture CIS cells, would produce only a gigawatt of cells, enough to provide just over one percent of the energy shortfall projected by the year 2000. Indium Phosphide (InP) shares the indium problem. Why should we spend money researching a technology that is so sorely limited by global resources? And what if, having devoted the planet's whole stock of indium to this effort, we then find some other, essential need for indium? Gallium arsenide (GaAs) is comprised of a rare metal and a poison. Cadmium telluride is similarly flawed; in this case the main problem may be the health risk: cadmium is known to accumulate in human bone, and we have no idea how much of it will find its way into our food chain. Does it make sense to repeat patterns of broadcasting pollution when a well-developed, nonpolluting resource is already available? Researchers working privately with more common and benign materials and strategies have found NREL to be unhelpful, and suggest that NREL's choices were based on the wishes of some politically well-connected scientists. Which opens the question: Does it make sense for us to ask our government to involve itself in research like this at all?

Manufactured amorphous or crystalline silicon-based PV cells are about as toxic as beach sand, but their production involves halide gases and aggressive toxics which must be very carefully handled and meticulously reclaimed and recycled to avoid environmental damage. PV manufacture consumes an enormous amount of energy; it will be a year before a module harvests as much energy from the sun as was embodied in it during its manufacture. Rigorous whole-life cost analysis shows that material extraction costs, substance risks, and production energy, if charged at fair, whole-cost prices, might add as much as 20 percent to the cost of each module we buy. This is a bargain, as externalities go, especially compared to burning nonrenewable fossil fuels. And at the end of their thirty-year useful cycle, when their output will have decreased so much it will be economical to replace the PVs with new modules, we should be able to recycle the ingredients very cost-effectively.

As we will see in the next chapter, families use varying amounts of electricity. To supply a household with photovoltaic power, you must first know how much power is required. Despite careful conservation, my house consumes an average of eleven kilowatt-hours of energy every day, and I generate barely a quarter of this with six photovoltaic modules. A photovoltaic module produces about 50 watt-hours of energy for every hour it spends in the full sun, and few sites are favored with more than six hours of full sun per day averaged over the year. A 2-kilowatt array, composed of forty modules, would provide adequate power for most of us if we eliminated waste.

Generators: Turning spin into flow

While PV is magical, conventional generation is quite simple. We have seen that electrons roam relatively freely in a conductor; they can be induced to move by exposing them to a moving magnet. A generator is simply a magnet and a coil of wire, one of which is held steady while the other is rotated, inducing a flow of electricity through the coil. When the coil's output is connected to a circuit, electricity flows.

Wind power is a perfect example of transforming a linear flow—wind— into rotational power using a rotor. Hydro power transforms linear stream flow into rotational power using a turbine. Dirtier techniques harness combustion directly or indirectly to piston-driveshaft arrangements which produce rotation. The rotational power is harnessed to a generator which produces a flow of electrons.

Wind power generally requires a minimum average wind speed of ten miles per hour (quite a windy site) to be economical. Wind and photovoltaic often make good partners, because wind usually blows when the sun is not shining.

Wind energy: Paul Gipe's story

Paul Gipe directs the Kern Wind Energy Association, a group of wind farm operators in California's Tehachapi Pass, between the southern end of the Central Valley and the Mojave Desert. He is an enthusiastic advocate of wind power, which he has seen work well in Europe. Considering its power and simplicity, Paul wonders why it is not more widely employed, and has devoted his working life to changing that situation.

There are some 5000 people in Tehachapi, and about an equal number of wind turbines. On a windy day, we generate enough electricity for Tehachapi, Mojave, the entire Antelope Valley, and we'll export electricity to Los Angeles as well.

Our production offsets generation and attendant pollution from nuclear-powered and fossil-fired plants in the Los Angeles basin. We estimate that our wind turbines offset enough carbon dioxide to fill a line of garbage trucks from Tehachapi to San Diego each year, a distance of more than two hundred miles.

The wind power plants here generate enough energy to meet the residential electrical needs of 500,000 energy-hungry Californians per year. If we had the same number of turbines in the Netherlands, they'd meet the needs of twice the number of people because a kilowatt-hour goes so much further in Europe.

Wind energy is a part of life here. The turbines are visible from town and there are 350 people here who work with wind energy in some way. In a small community, that makes a big difference. Everyone knows someone who works on a wind turbine or who works for a company that does.

It's too bad the town didn't develop the resource itself, and become a net exporter of electricity: it would have been a nice little municipal nest egg. One of the local political leaders suggested just that. But no one listened to him then.

Wind farms are a perfect way for communities to work toward energy self-sufficiency. There are several small municipal wind power plants in the United States and Europe. And a growing number of villages in developing countries use hybrid power systems, using both wind and solar energy, to meet their needs.

Wind development need not follow the pattern pioneered in California, where the turbines are found in large arrays. In Denmark two-thirds of the 3500 turbines there are installed individually or in small clusters. Many are owned by local cooperatives. They sell all their electricity to the utility, and use the revenues to offset the members' utility bills. Because they sell to the utility, they can site the turbine (or turbines) to best advantage, say, on a nearby hilltop, rather than placing the turbines in members' backyards.

Unfortunately, many of our rural-electric cooperatives here are not so foresighted. They're "cooperatives" in name only and often stop members from installing wind machines by forcing them to meet unreasonable requirements that long ago were proven unjustified.

Paul Gipe preparing to service a wind turbine (photo by Paul White).

It's because of the co-ops that I work with wind energy. I got started by tracking down old windchargers that were used on the Great Plains prior to the Rural Electrification Administration, which spawned the co-ops. I bought and sold windmills that were relegated to the junk pile after REA came through. The co-ops wouldn't allow ranchers to use both. They forced the ranchers to take down their windchargers. These junk machines were a boon to me during the mid-1970s.

I was studying the geo-hydrologic impact of strip mining in Montana over 18,000 square miles in the southeastern corner of the state. I kept coming across these abandoned windchargers. Finally I took the back seat out of my VW bug and began dickering with ranchers for what to them was just junk. By the end of the summer I had a tractor-trailer full of scrap windmill parts.

It was a wonderful experience. I'll never forget it. We renewables advocates were all so innocent then. We were out to save the world. Those windchargers fit perfectly with what I wanted to do.

I originally studied mechanical engineering at General Motors Institute of Technology in Michigan in the days when they had a dress code: no sideburns, no moustaches, a yes sir–no sir kind of place. After a couple of years of that I went back to Indiana. While there I began working with a student environmental group at Ball State in Muncie. We were fighting strip mining in the southern part of the state. I eventually got a degree in Natural Resources, but it was secondary to the environmental work.

One day I was testifying at a hearing when some canny old pol said, Son, those alternatives you talk about, they just don't work. At the time he was right, so I decided to put my life where my mouth was and help make renewables happen. You can't stop strip mining on dreams alone.

There were plenty of people already working on solar and it seemed, well, a little boring. After all, solar panels just sit there doing what they do without any fuss. There was also far less attention paid to wind's prospects than solar. It was the renewable underdog. Proponents would say, "We want solar energy," and then as an afterthought, "Oh yeah, wind energy too." That's still true today.

All in all wind energy simply appealed more to me. It was mechanical, and used the aeronautic arts. I grew up in the Sputnik era and was always fascinated by aircraft. Wind best fit my combined environmental and technical interests. I found operating wind turbines entrancing—still do. I felt I could make a contribution in wind.

So, I shipped those junk Montana windmills back East with every intent of refurbishing them for some working demonstration of renewable energy. While networking with environmental and energy groups in Pennsylvania, I called a dealer who rebuilt used windchargers. He showed up on my doorstep the next day, money in hand, wanting to buy my windchargers. I soon found that I'd become a wholesaler of junk windmills. And, as they say, "It's been downhill ever since."

When that lode played out in the mid-1970s, I began writing about wind energy. In 1988, the American Wind Energy Association named me the wind industry's man of the year. I guess they figured that anyone who has survived that long writing about wind energy deserves a medal.

Wind still offers tremendous promise. We estimate that there's enough resources for wind energy to easily meet 20 percent of the nation's electrical needs, even after excluding unsuitable areas like parks and wilderness areas. We often say that North Dakota alone is a potential Saudi Arabia of wind energy. If we get to just 10 percent during my professional life, I'll be happy. That's one hundred times what we have today. Technically we can do it. The question is, do we (the nation) have the will to do it?

The public seems willing, according to opinion polls. But the nature of utility regulation and entrenched interests make it problematic. Utility regulations, which vary from state to state, effectively limit what can be done. During the late 1970s and early 1980s California took a series of steps that ultimately resulted in the world's greatest concentration of solar, wind, and geothermal power.

Wind energy in California is one of renewable energy's greatest success stories. Within a decade we went from utter failure and public ridicule to generating 1 percent of the state's electricity from 1600 megawatts of wind turbines. This is a state with one-twelfth of the United State's population and the world's seventh largest economy. But the political winds changed. Except for stalwarts like Chuck Imbrecht at the California Energy Commission, California's renewable leadership was lost long ago. Europeans have since picked up the torch.

Denmark, which is one-tenth the size of California, expects wind to supply 10 percent of its electricity by 2000. By 2005 they plan to have as much on-line as California. The Netherlands, the most densely populated country outside Bangladesh, is equally ambitious. They've set an official policy goal of 1000 megawatts of wind generation by the year 2000 and double that by the year 2010. Two German states plan to install 1000 megawatts each by 2010. In contrast the United States has no goal at all. It's more than an embarrassment, it's a disgrace.

There are a few cases of enlightened utility regulation in the United States; Minnesota and Oklahoma come to mind. There, the owner of a wind turbine can sell power back to the utility at the retail rate, at least up to the amount of their own consumption. In most states, however, the utility will pay only the "avoided cost," typically 35 to 45 percent of the retail rate. Under these conditions it's difficult if not impossible to make wind pay, unless the wind machine is used solely to offset domestic consumption.

The reason the Danes and the Germans have been so successful is not because they have some well-funded, high-tech research program, but because they pay a fair price for wind-generated electricity. Danish utilities pay 85 percent of the retail rate, German utilities pay 90 percent. On top of that, wind generation in Denmark is excluded from the carbon tax, as it should be, and from sales tax. Most German states threw in an up-front subsidy as well.

If we got a fair price here in the United States, say 90 percent of the retail rate paid as a "Green" tariff, we could make wind work. If we can't make it work, then we should all find jobs building natural gas plants, because that's what the utilities are planning to build right now.

Wind development could follow three paths here in the United States: wind farms for bulk power; individual wind machines for homeowners, farmers, and businesses; and small wind turbines for off-the-gridders. Today, the off-the-grid market is doing well.

If the wind is available, small and what I call micro wind turbines are a far better buy than PV. But there's no beating a hybrid system using both PV and wind. You get the best of both worlds because they complement each other. In late summer when winds are light, solar reaches its peak. During the winter, the wind system picks up the load. Advances in inverter electronics, compact fluorescents, and other energy appliances have revolutionized life off-the-grid. As a rule of thumb, anyone living more than half a mile from a utility line will find building their own hybrid power system a better buy than bringing in utility power.

But wind doesn't have to mean only small turbines off-the-grid and thousands of utility-scale turbines in wind plants. There are other ways. As the Danes have shown us, we can install small- and medium-sized turbines for homeowners, farmers, and businesses who use as much electricity as they need, and then sell the rest. In Denmark, Germany, and the Netherlands, homeowners and farmers install the same size turbines as we install here in our wind plants.

Because cost-effectiveness often increases with the size of the turbine, the economics of individual turbines in Europe is more attractive than here. That's why single turbines are sprouting up all over the countryside. You can't drive down a road in Denmark and not see a turbine spinning in the distance. The same is true in northern Germany and in the northern provinces of the Netherlands. It's an impressive sight.

Again, what makes this possible is the price northern European utilities pay for wind-generated electricity. They pay enough to justify buying the most economic wind turbine available even if it will generate more electricity than needed by the owner. Everyone wins. The farmer benefits by getting a good buy on the turbine. Society benefits by getting clean energy for use elsewhere.

The best deal we get here is "net energy billing," and we only get this in a few states. This essentially allows a homeowner to bank any excess production with the utility at the retail rate by running their meter back wards. At the end of the month the utility balances any excess production against consumption. Any surplus above monthly consumption is paid the "avoided cost," typically only a few cents per kilowatt-hour.

Widespread use of net-energy billing would lead to a boom in small wind machines. In those states where it exists, such as Oklahoma, small wind turbines like the 10-kilowatt Bergey Excel make sense. In Minnesota, 35-kilowatt turbines have proven popular with farmers because of net-energy billing. There's a tremendous potential in the midwest for individual wind turbines if we can just tap the farm market.

That should be our vision, a mix of different kinds of wind turbines in different applications serving a variety of needs. One day we should be able to drive across the midwest, just as in Denmark today, and see a small wind turbine here, a cluster of bigger machines there, and further down the road a pleasing geometric array of hundreds more.

After twelve years of outright hostility towards renewables from Washington, D.C., we're hopeful that the nation's direction is about to change. The new administration offers some promise that new policies towards wind energy will be in the offing. We hope so. Wind has a lot to contribute. Wind energy could revitalize the midwest. If we can provide farmers there with a new cash crop like wind-generated electricity, they could stay on the land, tilling the soil like their forebears. Wind could provide manufacturing in the industrial heartland, and service jobs in thousands of struggling small communities. The technology exists, and the time seems right. We'll know by the mid-1990s if wind energy will blossom in the United States.

If you know the average wind speed your wind turbine will see, you can estimate its potential output by using this table. For example a Bergey 1500, which uses a rotor 3 meters (9.8 feet) in diameter, will generate 1900 kilowatt-hours per year at an off-the-grid site with an average speed of 10 mph. Its big brother, the 7-meter (23-foot) Bergey Excel, will generate five times as much: 10,000 kilowatt-hours per year. Because the wind resource is sporadic, it is not meaningful to speak of daily output; at times, an adequately sized plant will give you more than you can handle, and at times the spinner will stand becalmed.

ESTIMATED ANNUAL ENERGY OUTPUT

at Hub Height in thousand kilowatt-hours per year

Average Wind Speed in mph	*Rotor Diameter*							
	(m) 1	1.5	2	3	4	5	6	7
	(ft) 3.3	4.9	6.6	9.8	13.1	16.4	19.7	23.0
9	0.1	0.3	0.6	1.3	2.3	3.6	5.2	7.1
10	0.2	0.5	0.9	1.9	3.4	5.3	7.6	10.0
11	0.3	0.6	1.0	2.3	4.1	6.5	9.3	13.0

Falling Water: The best source

Falling water is the simplest and most cost-effective alternative energy source. If you have a good year-round source, your attitude toward electricity may change in surprising ways: you may be looking for ways to use extra energy. A site with excess hydro potential could burn the excess to heat water and space. An even better use would be to develop a neighborhood grid. Year-round hydro systems need less storage than solar or wind, enough to match daytime usage with twenty-four-hour production and cover times when the turbine and generator are being serviced.

A hydro site consists of a *forebay*, from which water is directed into the *penstock* or pipe which carries it down to the generator and back into the stream. The water, under considerable pressure due to the *head*, or vertical drop, runs through one or more nozzles and strikes a *turbine*, which converts the flow into rotational energy that drives a generator, often an automobile alternator. The critical factors that determine a hydro system's success are flow and head, variables that favor a wet, mountanous site. *Flow* means a quantity of water, usually measured in gallons per minute (GPM), often surprisingly small, which can be impounded and made to drop a goodly distance. Head need not be a dramatic drop, as in a waterfall, but is the vertical distance between the intake at the forebay and the generator site; if the horizontal distance, called the *run*, is long, the flow will be decreased, but a ratio of four feet of run to one foot of head is enough of a slope. Assuming that a small household requires 500 watt-hours a day, a flow of ten gallons per minute with sixty feet of head will provide an abundance. My Caspar household would get by on fifty gallons per minute falling one hundred feet, which is quite a lot of water.

The relationship between flow and head is complex (see table 3 below). Micro-hydro uses relatively small amounts of water falling a considerable distance; serious energy production starts when the head gets to about fifty feet. Low-head, high-flow systems, usually involving river water with only a few feet of head and a flow reckoned in cubic feet per second—a lot of water!—can be incredibly powerful, but necessitate a dam or diversion and large equipment. Such potential is very site-specific, and requires expert planning.

Every hydro project presents a unique combination of water source and topography, and requires more careful surveying and planning than an equally productive wind or photovoltaic installation. Developing a hydro resource is often complicated by environmental concerns, by neighbors, and by local regulatory agencies. Water quality, fish migration, and the effect on the portion of the streambed from which water is diverted are all legitimate issues that should be carefully studied and addressed. In fact,

ESTIMATED MICRO-HYDROELECTRIC POTENTIAL
in Watts of Output

	Head (in feet)						
Flow in GPM	25	50	75	100	200	300	600
3	—	—	—	—	40	70	150
6	—	—	10	20	100	150	300
10	—	15	45	75	180	275	550
15	—	50	85	120	260	400	800
20	25	75	125	190	375	550	1100
30	50	125	200	285	580	800	1500
50	115	230	350	500	800	1200	++
100	200	425	625	850	1500	++	++

hydro projects do not use water, they merely borrow it for a few moments, so a hydro project will in no way diminish the water resource; infringing upon the water rights of others is seldom an actual problem except in the rare case where water is diverted from one watercourse to another. And yet, regulatory agencies live to regulate, and private power generation makes them nervous. For all these reasons, many micro-hydro systems are undeclared, and micro-hydro is probably more common than we think. I visited one system in New England where the operator, a farmer whose family had held the land for four generations, installed his wheel as a defiant act of independence, and was thrilled to have a secret source. More power to him!

Most hydro sites are seasonal, producing well during the stormy months when PVs are unproductive, which makes the two sources good mates. Although studies suggest that only a tenth of our nation's hydro potential has been developed, ideal hydro sites are as rare as they are perfectly suited to independent power production. In many western states, despite regulatory resistance, hydro enjoys a very favored treatment: an undeveloped hydro site can be developed against its owner's will in a process quite similar to exercising eminent domain, because the source is recognized to be of greater community importance than mere ownership; it is understood that water does not belong to the owner of the property it crosses.

Hybrid Systems: Multiple energy sources

It is a rare site where a single sustainable source will provide adequate power through all weather conditions, and so use of multiple sources is necessary. If you are lucky enough to have access to wind or hydro resources, you will find they often complement the photovoltaic source well, since water flows and wind blows when the sun is not shining. Photovoltaic and wind generation seldom use the whole available resource, and so their production can be increased simply by installing more modules or another wind-spinner or two; unfortunately, both still produce relatively expensive electricity, and so we are tempted to look for a cheaper source. Most existing homes are on-the-grid, and that is a clean and economical source compared, for example, to backyard fossil-fueled generation.

Utilities do much of their generating of electricity by burning nonrenewable fuels to boil water into high-temperature, high-pressure, superheated steam, with which steam-turbine generators are driven. Fossil-fuel-fired generators work well; the bigger they are the more efficient and cleaner they can be. Smaller units that use internal combustion or diesel engines to drive generators can provide good back-up during power outages in hospitals and other places requiring absolutely reliable power, and are common in off-the-grid systems. Unfortunately, these units convert fuel to electricity inefficiently and are environmentally dirty; smaller, in this case, is uglier.

Fossil-fuel-powered generation is the last, worst choice for back-up power from the standpoint of cost and pollution. When pollution is considered, propane is the fuel of choice; where operating cost is the determining factor, diesel is best, and propane is a close second; for small, locally available, consumer-proven machinery and fuel convenience, gasoline-powered generators are the easiest (and dirtiest and least efficient) choice. At the most successful installations, where householders manage their energy usage realistically and carefully within the context of sustainable energy production, generator use decreases with time as they learn to live within their renewable energy budget. Often the generator stands unused for months, and is started a half a dozen times a year to keep it limber, to power a large tool, or to recharge the batteries after a lengthy cloudy spell during that awkward time in early fall after the sunny season but before the storms come with enough regularity to recharge the stream or set the windmill spinning.

I visited a farm in a wooded valley with a strong year-round stream where a huge 25-kilowatt diesel generator and industrial-strength battery bank had been installed by the previous owner to allow for a no-load-spared, all-electric lifestyle. The generator, the size of a small tractor, ran an average of five hours a day, and the day-to-day costs of electricity compared favorably with grid power. However, overall costs, including original equipment and periodic major engine-generator maintenance, were staggering. In practice, the arrangement reproduced in miniature what is wrong with fossil-fuel electricity production: the generator, which started automatically when battery state of charge reached a setpoint (which often happened in the pre-dawn cold), would kick in to recharge the batteries, and would fill the pristine valley with its ghastly throbbing and a miasma of local smog that lingered for hours after the generator shut down, trapped by the protected little valley's natural inversion layer. The new owner, who understands the value of alternative and nonpolluting energy, can hardly wait to pull the plug on the diesel and harness the stream and the sun.

Power's Dark Underbelly

Dead dinosaur fumes! It is not hard to imagine the harm that comes to our bodies when we breathe the leftovers of burnt fossil fuels. Nature went to so much trouble to get fossil fuels safely buried, but now our race is in such a hurry to dig them up and burn them dirtily! Too often in my travels, I found the generator sited absent-mindedly, like a filthy secret, in the lee of the house where its exhaust quickly permeated the living space. The best installations acknowledge the presence and need for the fuel-burning plant, and cede to it a shed of its own, sufficiently downwind to abate, at least locally, the noise and stench.

As appropriate technology becomes part of our everyday lives, it is hard to remember that residential electricity is a twentieth century idea, as new as the dependent home. Like most newfangled ideas, we treat it cautiously at first—as though knowing that, in time, we might find that it has hidden hazards and is not such a good idea after all. But caution subsides with familiarity, and now we see that we have willingly entangled ourselves in a web of wire.

The wires strung across our countryside look like the kind of temporary work I do in my house for Christmas tree lights or when I'm testing something new. If the device works and I like what it does, I figure out a way to install it permanently so that its support mechanisms do not obtrude on my aesthetic consciousness. Can we agree that electricity is a good idea, and we would like to keep it around? Good, so now let us put it in the walls and under the floor where it belongs. Then we can begin to handle it in a way that doesn't make us constantly conscious of its haphazardness. After almost a century of impermanence, planners and customers in upscale communities are forcing the utilities to put the rat's-nest of utility lines underground where they are safe and invisible.

Apart from their offense to the eye, can power lines be good for the plants and animals that live near them? We are learning about the possible results of the distribution scheme—stray current, electromagnetic pollution, incidental microwave radiation. This is doubly offensive, because such radiation represents distribution loss and grid inefficiency. To many of us, the grid seems an elaborate, unnecessary, poorly designed scheme; increasingly, where reliable, uninterruptible power is required, renewables are the source of choice.

One standard argument against photovoltaic (PV) power advanced by power interests is the assertion, based on the monstrous consumption of the 1950s All-Electric Home, that producing enough power for a home requires so many hundred square meters of PV modules. By curtailing waste, which amounts to more than half our electrical use, we know that an average home's energy use can be harvested by modules covering an area about half the size of that home's roof. After thoroughly purging wasteful energy equipment from our homes, a program of covering residential roofs with grid-interconnected modules could supply more than enough power, using currently available technology, to satisfy all residential and office requirements; existing capacity could then easily accomodate growing requirements in the manufacturing sector. This solution, and the cosmetic improvement of undergrounding utility lines in every community, promises a functional, long-lasting, and appropriate relationship between people, houses, and electricity.

The appurtenances of twentieth century technology, looping power lines, satellite dishes, and solar trackers, do not settle easily into the peaceful dignity of a New England farmhouse foursquare on its land, or the rugged beauty of California' s wild north coast. I wondered, as I visited independent homesteaders across the country, how these high-tech gadgets would look if they had they been developed at the same time the saltbox house was inventing itself. I got a hint from the wonderfully ornate, wrought iron, glass-balled lightning arresters on some New England barns: humorous and functional, these finials remind us that invisible powers like electricity are not far removed from magic.

CHAPTER 4

HOME UTILITIES

TO TAKE RESPONSIBILITY FOR OUR OWN POWER, WE MUST MANAGE AND MAINTAIN energy generation equipment, and the hardware that makes up a system: transmission, storage, distribution, and metering. To do a good job, we must have an intelligent load management plan.

The starting point for such a plan, whether for a small remote home or for a utility monopoly serving millions, is a reasonable estimate of the power required. How much power is needed, and when is it needed? Utility dispatchers use this information to predict load and provide generation capacity to satisfy the daily and seasonal demand patterns of their customers. In a household grid, where you and your family are the customers, and power consumption is limited by your system's generative and storage capacity, a realistic load management plan will reconcile capacity with your requirements. This may simply mean that two big loads—the washing machine and the iron—cannot be used simultaneously. It may mean that, at times when generation is at a low ebb—for example, if your power comes from photovoltaic modules and the weather has been cloudy for a few days—that certain activities will be deferred so that basic power needs like lighting can be met. At the utility level, where customers have come to expect as much power as they can use whenever they wish to use it, dispatchers anticipate demand and try to use marketing and power management tools to adjust loads to local or purchased power availability. These tools include controlled brown-outs (utility voltages are lowered, putting some equipment, especially motors, at risk and decreasing their lives), rolling outages (neighborhood- or town-sized areas of service are shut down temporarily), and time-of-day metering (customers pay a premium for energy consumed during peak-demand periods, and pay less for energy used at other times). All of these intrude on domestic activities and therefore advance the argument for some degree of energy independence.

How much power do you need? There are three answers, two simple and one good. The simplest, most pessimistic answer is that we always need more power than we can possibly produce sustainably and locally. This answer assumes that we insist on using any amount of energy at any time, without considering load management, and unless we learn to work efficiently and within our energy budget, we will need the grid.

If you buy electricity from a utility, the second simple answer can be found by reviewing your monthly electric bills and calculating a seasonal daily average. Normal households use between 500 and 50,000 watt-hours per day, which is a very wide range; my house in Caspar consumes, according to the second simple method of calculation, between 6 and 11 kilowatt-hours per day, half of it used by computers. On a good day, a quarter of my electricity is generated photovoltaically.

Figure 15.

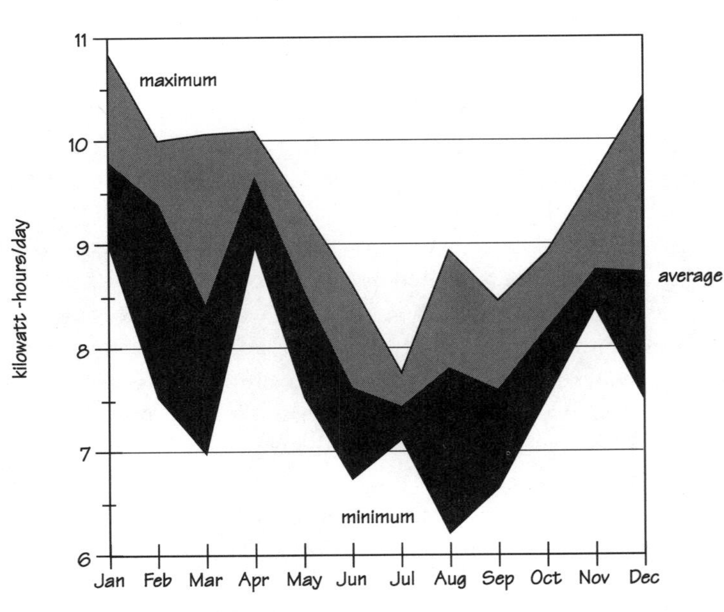

Home Energy System Overview

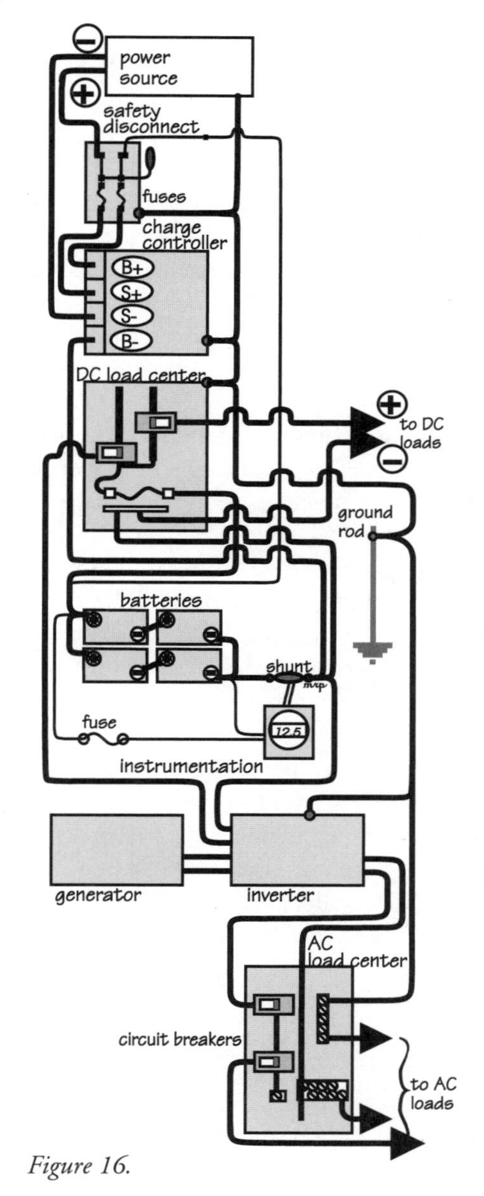

Figure 16.

The energy source for an independent home power system can be photovoltaic, wind, water, biomass, or some other form of generation; in the accompanying diagram I have left it a blank box. Whatever the source, there will be times when we will wish to disconnect it from the house's wiring, and the Safety Disconnect accomplishes this. The S-curved devices in the diagram represent fuses, which are the first line of overcurrent protection. When too much current flows through one, it quickly heats up and "fuses," thereby breaking the electrical circuit. A fuse is a one-time device; once it fuses, it must be replaced.

The charge controller does what its name suggests; it controls the amount of electricity from the source that is applied to charging the battery bank, based on the state of charge of the batteries. The DC load center allows control of each low-voltage circuit. Each of the DC branch circuits is connected to the load center through a circuit breaker, which is an electrical analog of a fuse. When too much current flows through a circuit breaker, it "trips" and interrupts the circuit; when it has cooled down, and the overcurrent condition has been corrected, it may be reset by switching it off, then back on. Two wires conduct electricity between any two components in the power system, because electricity only flows in circuits; imagine a stream of electrons flowing away from the minus side and toward the plus side. For safety, a third wire connects all metal frames and boxes to each other and to a ground rod driven into the earth.

Even in a low-voltage DC system, safety must be the guiding principle. Note that even the meter is fused and grounded.

Under normal conditions, when adequate energy is flowing from the source or is stored in the battery, the inverter changes the low-voltage DC power into house current, 110-volt alternating current (AC). In this system, the inverter can accept electricity from the generator; when the generator is providing power, the inverter uses electricity from the generator to recharge the batteries, and isolates the low-voltage system from the irregularities in the AC current. The AC load center provides convenient connections and circuit-breaker overcurrent protection for AC circuits.

Planning domestic power: Felicia and Charlie Cowden's story

Charlie and Felicia Cowden live off-the-grid in an house they built on Kauai's North Shore. In the aftermath of Hurricane Iniki, which bludgeoned Kauai in September 1992, they helped their neighbors get their power back. In sunnier times, they run the Hanalei Surf Company. As you will see in their story, the move from the grid required learning and adaptation, but the conclusions they have drawn will work for anyone who uses electricity.

Felicia and Charlie Cowden.

Felicia: The biggest stumbling block on the way to power independence is our culture's custom of wasting incredible amounts of energy. We waste more than we use. You can't simply buy energy independence: you have to make a brain investment, and learn some things.

We started moving toward conservation when we were living on the grid in Princeville by replacing incandescent lights with compact fluorescents.

Charlie: When we were deciding to buy our property and build, whether to pay the money to trench and connect to the power lines at the highway or go solar, I didn't know anything about alternative energy sources. So I asked my contractor, should I do it? He didn't know much about it either, but he said, "It's the future, man, you gotta do it."

Felicia: We shopped around and bought parts from different sources; sometimes they weren't even meant to work together. Because we got different parts from different people, we had some problems getting people to help us put things together right.

Charlie: So I drew a picture of my charge controller and sent it off to the guys that made it and said, hey, I'm thinking about putting this stuff together like this, what do you think? And the same with the batteries and inverter. I did it the way they told me was best.

Felicia: We throttled our energy use way back when we moved to our new house. We were limited financially, and could only get six panels. We figured we would need more, but during the sunshine months, we're full by noon, and we haven't given anything up. We just changed our energy habits.

When we first moved in, it was like living in a science experiment. It soon became very apparent that there were things we hadn't thought of before. Living in my house has made me much more conscious of consumption, so when I visit other people, I see incandescent lights, and can't help but notice all the things being done inefficiently. It feels sinful! Solar power's biggest gift to the environment is showing people that it's possible to live well without being wasteful. Since moving in, I've become an energy evangelist, always trying to get more people into conservation.

There are so many things we wish we'd known! Nobody told us about phantom loads, things that use electricity whenever they're plugged in, like the clock on a microwave. Four little clock radios can use up everything our panels produce in a day. We use battery operated clocks with rechargeable batteries. The real energy criminals of the small appliance world are the remote products that require chargers, like electric toothbrushes and cordless razors. They draw substantial current to charge their batteries, then when the batteries are full, they continue to trickle energy. That's a lot of burnt dead dinosaurs for a few minutes of minor convenience, and we all have to breathe the smoke. Much better to use appliances that plug in directly and only burn power when they're in use. If push comes to shove, you can always brush your teeth yourself.

When we go to bed, it's usually really quiet, but if I still hear the inverter humming, I go look at the meter. I've learned my phantom loads: if it says 2 amps it's the microwave, 4 amps is the TV and VCR, and 8 amps is the sine wave filter for the stereo.

Charlie: After Iniki, I was helping out an electrician friend installing gensets—generators, you know?—and when we'd get ready to start the engine, we'd tell the owner, "Go in and turn everything off, okay?" and pretty quick he'd come out and say, "Okay !" So we'd fire up the generator, and immediately see 100 amps being drawn out of the system. We'd figure there was some kind of short or something wrong because of the storm, so we'd shut down and go looking for the problem. We'd find a dustbuster drawing 30 amps over here, and a little convenience fridge drawing another 30 over there . . .

In our house we put our phantom loads in power strips, but it would be better to put switched outlets in the wall.

Felicia: A perfect example: a recessed microwave with the outlet behind it, where you can't reach it to unplug its phantom load. Lots of times, people want to install their microwaves that way. If you put a wall switch on it, you can turn it off when it's not needed.

If I leave all our phantom loads—the VCR, stereo, TV, microwave, and adding machine—plugged in, even if they are turned off, they consume more electricity than my panels produce. If I can turn their outlets off, I produce more power than I need. So you don't have to do austerity things, you just have to be careful not to waste.

We learned a lot since we built our house, most of it what the power company calls load management or demand side management. I can't run the stereo when I'm doing a wash, because I chose a power-hog double agitator washing machine. Now, I'd buy a front loader.

Charlie: You've got to make small amounts of electricity do your work. I saw a front loading washer in the store, and I thought they'd made a mistake because the energy tag said it used a tenth the power of the top loaders.

Felicia: It's all in the torque. You've got to understand the difference between inductive loads, like motors and little transformers, and resistive loads, like lights. Inductive loads are real power hungry, especially when they start. For example, you learn not to run the vacuum cleaner when the washer's running. The garbage disposal puts a big strain on our system. . .

Charlie: . . . and the television picture gets real small.

Felicia: If I wake up in the morning and see the batteries aren't depleted, I look for high demand things to do, like washing and vacuuming. Load management depends on weather. If it's cloudy, and the batteries are depleted, I think about conservation.

Charlie: At first, demand side management is a hard thing to get people to think about. They still want to do things as if they had infinite power, and they have equipment already . . .

Felicia: . . . their energy hog refrigerator, which would take twenty-four modules by itself . . .

Charlie: We tell them they can't take that equipment along, they need to start out energy-thrifty.

Felicia: I start out with Kauai Electric's energy pie, showing what the average household uses, and tell them that you can do all that with a $300,000 solar installation, so now let's start managing. Solar power is an excellent application in the tropics, because we don't have to heat or cool space. We install solar water heating. We substitute a Sun Frost or a Sunshine Coldmizer for a conventional fridge. Manage lighting and phantom loads. We personally reduced our power consumption by 95 percent. That's substantial.

Another thing I wish we'd done: our inverter puts out a modified square wave, not the sine wave that most appliances are used to. Before buying an appliance, find out from the manufacturer if it needs a pure sine wave to work right.

Charlie: Felicia bought a stereo, and when we plugged it in, the lights went on, but no sound came out. I went, okay, no tunes, and turned on the TV, and the stereo started up! So I turned off the TV, and the stereo stopped. How's that? I checked with the stereo techs, and they told me the TV had capacitance that smoothed out the power. I called the inverter techs, and they said, "Capacitance, hunh? We can send you some of that!" and they sent us this big capacitor which I put in the system. It worked okay, only seemed to draw 50 watts, but when I went to check the inverter, it was really hot. Whatever kind of 50 watts that capacitor was using, the inverter didn't like, so I disconnected it. Finally we got a sine wave filter that uses lots of electricity, but lets us use our stereo.

People think they've got to make sacrifices, give up TV and live in the dark. That's not necessary, they just have to pay attention. Take glow plugs for example . . .

Felicia: Yes. When you're buying a gas stove, it's a big clue if you have to plug it in.

Charlie: I think it's illegal to sell a stove with pilot lights in some places, because they use so much gas, but the glow plugs burn an awful lot of energy.

Felicia: We like to put inverters and batteries in an outside building. Our inverter is in the bathroom, and if you're in there and the toaster's down, you hear the inverter whining. It's putting out a worrisome amount of electromagnetic radiation.

Charlie: So when the washer is going, we scatter.

Felicia: When you don't meter your system, it's like having a car without any gauges, no odometer, no gas gauge.

Charlie: So you wouldn't know when to go to the gas station. I tell people that if you've got a PV system, you've got to have meters, too. You can't make the system work right without the meter. I went to a house where the guy had been living off-the-grid for fifteen years, and all that time the system has been growing, adding batteries, modules, and circuits without labels, mostly running off his generator. But he doesn't understand why he runs his generator so much. He's finally redoing his system, and adding an amp-hour meter. It will be a born-again system.

Felicia: Whenever we go into a house without a meter, we find that even after years and years, the people don't understand their systems. When guests come to our house or people come up to work, they are fascinated by our system, the fact that we have all the electrical things we need. When their friends come by, they like to play with the meter, show what the system is doing. It's instructive and fun.

It helps if the system is shipshape, like it was meant to be part of the house. When people see batteries in a corner and all sorts of cables, it makes an image of alternative energy as a subculture lifestyle. That's not necessary or even true. It works better if it's clean.

It's worthwhile to check local requirements. Here, you have to get past the inspector to get fire insurance, and you can't get financing without fire insurance.

Charlie: You've got to find out the rules—and there are so many rules. You must be careful to comply with building codes, or if there's a fire, the insurance company will find a way to deny your claim. I'd be scared to death to work on an unsafe, non-code system. If you don't use the right practices, and something goes wrong . . .

Felicia: When we talk about energy self-sufficiency, we try to get people looking away from the money issues. Maybe you save money going off-the-grid, but there are more important things. When you go to sleep at night, there's a zero electromagnetic field. And think of the tons of stuff you aren't putting into the atmosphere. It's a bonus if you save money.

People on the North Shore have a wonderful attitude about work, which taught me a lot about pride. Here, if you ask somebody what he does, he's likely to say, "I surf." My friends here are carpenters, plumbers, electricians, but they are all surfers, really. With them, I want to be me, not what I used to be, or what I do to make money. In fact now, that's how I answer that question. Somebody says, "What do you do?" and I ask, "You mean, to make money?" It makes them think, sometimes, that there might be more important things to do.

Storing Energy

When sources produce energy at the same time that loads require energy, the match is perfect. This is surprisingly common: when the sun shines, algae grows and pool filtering is required; when the sun strikes the solar hot water panels and the water is heated, a photovoltaically powered pump provides timely circulation for an active hot water system. In both cases, a PV module directly driving a circulating pump matches the source to the demand, and eliminates the need for other controls. Year-round hydro systems may produce so much full time power that little or no storage will be required, and the problem instead becomes finding uses for the excess.

Most often, source and demand do not match. If photovoltaic power is required for nighttime lighting, but the source is active only during the day, a means of storing energy is required. The grid has allowed us to ignore our sources when using energy, and until we correct this mental habit, we will need to use stored energy from fossil fuels.

The primary reason to store photovoltaic energy is because the sun doesn't always shine when we want power. The sunniest sites enjoy more than three hundred thirty days a year without cloud cover; other sites are lucky to have two hundred. The problem comes, even to those in sunny sites, when storm clouds cover the sun for several days at a time. In Caspar and at the Rocky Mountain Institute, these periods have been known to last almost forty days and forty nights. Unless back-up is available, storage must anticipate this worst case. It makes sense to store at least enough energy to weather the average storm and minimize the use of fossil-fueled back-up. The number of days a system can run on storage alone are counted as the system's days of autonomy.

Energy can be stored in three ways: mechanically, physically, and chemically. Mechanical storage is most familiar in a wind-up clock, where rotational energy is stored in a spring. Another mechanical technique for energy storage is through compression of a gas, the commonest gas being the air we breathe. Compressed air is an extremely clean, clear power source which can be used to power hand tools and small turbines like a dentist's drill. Compression inevitably produces heat: the molecules of gas are forced into closer contact with their neighbors, and although each particle has no additional energy, there are more particles within the tank and therefore more energy bouncing off the tank, creating pressure and heating it. The technique is inherently inefficient: as soon as compression ceases and the tank is allowed to cool toward equilibrium, some of this energy is lost.

In a fabulous example of physical energy storage, PG&E, Northern California's huge utility company, takes advantage of the fact that an electric motor can be designed so that its shaft will spin if electricity is applied to it, and will generate electricity if its shaft is spun. Using excess electricity during off-peak hours, these motor generators pump water from a river at a lower elevation to a higher lake. During peak hours, when more electricity is needed, the water is allowed to flow back down, spinning the motor generators and generating electricity. This technique is also inefficient, but the inefficiency here is irrelevant, because excess power is being stored then converted into power when needed. A more down-home example of good storage technique involves photovoltaically powered pumping of water from a well into a water tower during the daylight hours, then enjoying the gravity-fed water pressure at any time of day. The storage tank must be big enough to supply household water during the times without sun; the pump must be large enough, and supplied with enough power, to fill the tank during sunny times.

Chemical storage of energy involves a reversible chemical reaction. Petrochemicals and their younger relation biomass are slow forms of chemical storage in which sunlight, nutrients, and water are patiently transmuted into an embodiment of energy: vegetation. Fossil fuels have been further refined by time, temperature, and pressure so their energy has become quite dense and gooey.

Many chemical reactions are candidates for chemical storage, and every one of them has some problems. Wet batteries are most often found in automobiles. Dry batteries, found in flashlights and other portable electric devices, work in a similar way, except that only certain types can be recharged.

There is no such thing as a perfect battery; a whole-life cost analysis of

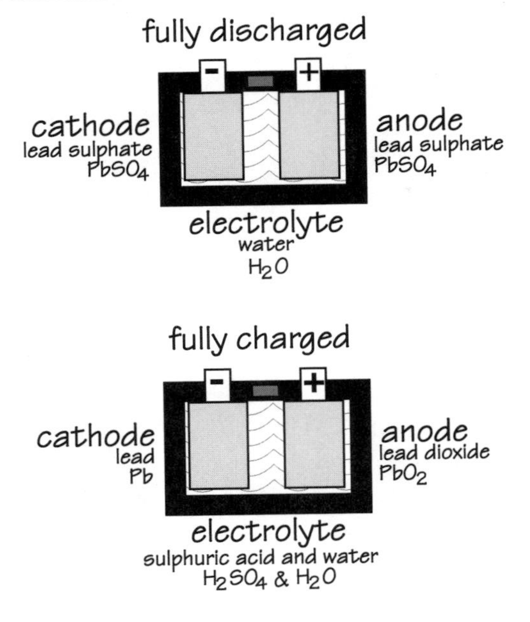

Figure 17. Fully spent, a battery's electrolyte has turned to water, all ions having been taken up by the cathode and anode. At the peak of readiness, a battery's plates have given up the ions to the electrolyte, which is now a corrosive mixture of water and sulphuric acid.

most battery technologies leaves an honest evaluator depressed. The most widely used batteries are called *flooded lead acid batteries.* Consider the creation costs: batteries are extractive nightmares containing large amounts of embodied energy (the energy required to manufacture them). They are hazardous and heavy, therefore expensive to transport. Once in service, batteries are dangerous to handle, giving off explosive and corrosive vapors, and requiring constant tending. Now consider the disposal cost: in a residential home system battery lifespan is between a year and twenty-five years (depending on type, treatment, and many other factors) and batteries are classified as toxic waste at the end of their lives, with only a small potential for reclamation and recycling. Even so, with all their unendearing traits, they are at the very center of a home energy system, and we had better love them. For the time being, they are the best game in town. I need hardly add that the invention of a better energy storage strategy will be greeted with delirium by independent energy producers.

We measure the efficiency of any storage device in terms of how much of the charging energy can be recovered, and by this measure, flooded lead acid batteries perform well.

The Battery Bank

Despite their problems, batteries are the best energy storage devices currently available, and so every independent home has an imposing battery installation—impressive at least in weight, if nothing else. Because of their weight, their noxious fumes, and their aggressive liquids, batteries deserve a place of their own, removed from the living space but close enough to minimize voltage drop, because direct current (DC) doesn't travel well.

Access should be easy, because batteries will fail and replacement will be required. It is a good idea to provide space for twice as many batteries as you think you will need; experience shows that you will underestimate the system size required, and your requirements will grow.

The best attitude to hold about batteries, I have found, is to treat them the way you might a pack of difficult pets. They respond well to love, but will find ways to punish you if you neglect them. They are choosey about how, and how much, you charge them, and fail quickly when overcharged repeatedly (although, perversely, they do like to be overcharged occasionally, which battery mavens call *equalizing*); nor do they like to be discharged quickly or completely. They do not like extremes of any kind. They perform best at a constant temperature, are sluggish if cold, and fail prematurely in heat. They like exercise, to have electricity cycled through them, but prefer to be kept nearly fully charged, which is called *floated*. If left unused, they lose charge and fail, usually irretrievably; if left unwatered, they fail. At different times during the battery cycle, they give off hydrogen, hydrogen sulfide, and oxygen gases; the first two are explosives and the latter encourages explosions. Under some circumstances batteries also give off a mist of sulphuric acid which corrodes all metals with a special preference for places where dissimilar metals touch each other.

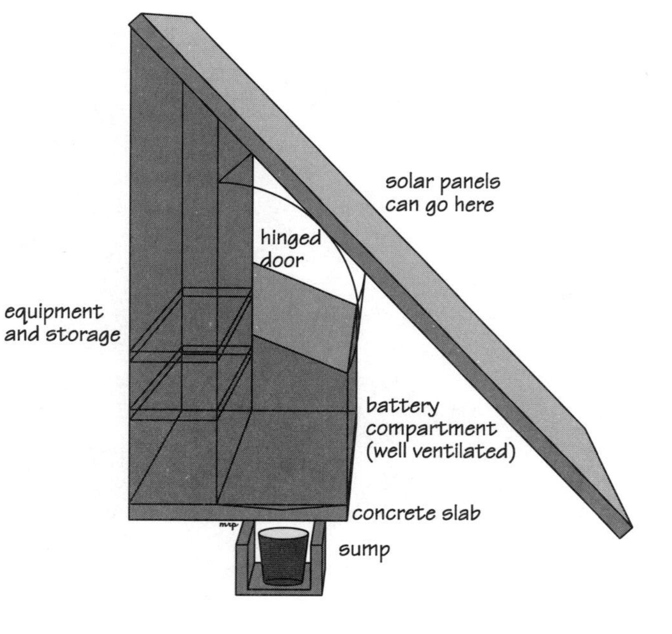

Figure 18. A well-ventilated battery cabinet's hinged door swings under a canopy of solar panels. The slab floor of the battery cabinet is sloped to a sump deep enough for a plastic toxic-spill bucket. The walls of the battery cabinet bolt together firmly, but may be disassembled so batteries need not be lifted. On the side away from the sun, the balance-of-system equipment, including the charge controller, inverter, and battery maintenance gear have their own cabinets.

By keeping the batteries clean, well-watered, with all their metal surfaces greased (any grease will do) or otherwise protected from corrosive mists, the negative attributes can be kept just under control. Batteries behave spectacularly in a fire, and are considered Class B explosives by the materials handling industry, so electrical inspectors in some jurisdictions require that they be kept in two- or four-hour fire resistant enclosures. Despite the greatest precautions, spills will occur, and so allowances should be made for containment, removal, and rinsing without shifting the batteries. Batteries do not get lighter with age, and battery acid is aggressive, so the foundation on which the batteries rest should be strong enough to be unaffected by moisture and acid; concrete is good, while wood must be protected with an impermeable, acid resistant covering. When designing a battery enclosure, consider all these factors.

It seems silly to mention it, but I have made this mistake myself: the battery enclosure must be ready before the batteries arrive, because these are unpleasant heavy puppies and you do not want to move them oftener than you have to.

Taming the horses: Dick Britt's Story

Dick Britt lives on and works an eighteen-hundred acre cattle spread in the high country above Snowmass, Colorado. Robert Sardinsky, more generally known as Sardo, is a part-time cowboy who works with Dick. (We will hear more from Sardo later in this chapter.) While Sardo had experience with alternative energy systems, Dick came to them without preparation; once he found the reins, though, he has taken full control.

Dick: We run about three hundred yearlings here in summer. Winter is calmer: tending the horses and mules, fixing the snowmobile, keeping the road open if I can, and tinkering with the power system. Summer's hard; you're either working fence, irrigating, doctoring cattle, or you're too tired to get out of bed. In summer, my twins, Dottie and Darby, are my cowhands. They've been here ever since they were eight years old.

Mostly, in winter, it's just me and the animals, but they make it worthwhile. Kitty-boy here showed up a couple years ago; he's never been inside, but he brings mice to me at the door. Betsy, my mule, she's thirteen years old: she got her eyes ruined travelling back and forth to Utah in an open trailer, so I keep doctoring her eyes. My next project is something to keep a hole open through the ice in the watering trough. Any ideas? See, we're always looking for new things to do.

When we first started, we just wanted lights and TV in the cabin. Sardo set me up with four solar modules and eight batteries seven years ago. Before that, we alligator-clipped our TV onto a jeep we parked up against the cabin. We'd run the battery down watching something, then jumpstart the jeep, rolling down the road, drive to the bottom and back up, and be able to watch a couple more hours.

Well, the first panels and batteries worked good, so we added some more modules, and an inverter so we had TV in the cabin and the shop.

We get plenty of snow here. In summertime people look at the panels mounted way up in the air, and wonder why we built them so tall, but they're just above snow level now. Beats having to shovel them off.

The guys that brought the second round of material must have been in a hurry, because they pretty much just threw the gear off their truck and beat it out of here. Nobody adjusted things, so we found a local guy, and with a little technical help, got everything right. I learned how to equalize the batteries, and keep it all going. We keep moving things around to make the system better. Batteries are like a team of horses: you want them all to pull together, but some are stronger, and you've got to match them so they work well.

Sardo: You can see that for our first installation, we took the easy route and surface-mounted the wires. On the second go we felt more confident, and Dick welded up a three-foot drill bit so we could thread through the wall, and we got a better looking installation.

Dick: When I got here, it was all like that. We had an old well, just a hole dug in a spring, a wooden box, any old kind of pipe lashed and taped together. A few years ago, we went back and did it right, and now we've got good water in all three buildings and the corrals.

Dick Britt regards the weather on a gray Colorado Sunday. The array has six feet of snow below it.

Wintertime, it can get real cold. I used to work nights, running a snow cat grooming runs at Buttermilk, and I'd come home and get the fire going. During the day it'd never get above freezing in here. So Connie, the owner, said we have to fix up a better bunkhouse.

Connie made all the plans and arrangements, and it was started in 1991, and was mostly finished during the summer of 1992. Mostly, the new bunkhouse got built well, but I'm pretty upset about how some of the systems were set up, kind of helter skelter, like there was no plan.

Sardo: The radiant heating system in the floor was a good idea, and they threw a lot of money at it, but it wasn't well-designed. Look at the pipe runs between the solar hot water heater, the back-up heater, and the circulating pump: it looks like the plumber from hell installed them. It will be a nightmare to replace anything in there. Designers need to think beyond building systems to how to maintain the systems down the road. Installing something under money pressure, skimping on disconnects, unions, whatever, makes maintenance real tough.

Dick: Part of the problem is in the way things grew. We have three propane tanks, two old ones on the cabin and a new, high-pressure one with high-tech fittings on the new bunkhouse, so when I ran out of gas in the new tank after three months, I was lucky I wasn't snowed in yet. I'm almost out again, and you saw the road . . .

Sardo: It would be better if there was a single large tank. That's the problem when systems grow without anyone thinking through the interconnections. You end up with several related subsystems that may or may not interconnect well—more often not. And there are two separate PV arrays, battery banks, and inverters in the two buildings, but it would be better to centralize systems. It makes a lot of extra work, maintaining separate systems.

Practically every project I have been involved with has grown beyond the wildest expectations. But you can't anticipate your ultimate needs and desires, let alone build a system that will meet them from the beginning. You can, and should, overbuild the infrastructure: mounts, interconnects, disconnects, metering. It saves a lot of work and money down the road. That way, you can add on, rather than start over. Sometimes you have to step back, then start over again from a higher level. Otherwise, you get a collection like this, where at best you burn excessive fuel to make it all work. The back-up generator was only tied into the cabin's electrical system, and the bunkhouse system had no back-up, so when it ran out of juice in the blizzard, Dick had to run extension cords to keep it going.

Commissioning is another part of the problem. A system is not going to work well without all the finishing touches tended to, setpoints adjusted, everything labelled, maintenance instructions provided, the people trained in how to look after the system. Those things have to be agreed upon as part of delivery, or they don't happen, and the system never works well.

Despite the problems, here's a super-insulated bunkhouse, with low-E glass, and open, airy space that works much better, Dick will have to agree, than what came before.

Dick: Each part of the system has its own personality. We found this guy, Kent, who'd been on atomic submarines and knew a lot about batteries and the rest, and could help me understand what was going on. I finally got control of it, going in the right direction. But it took two years looking to find the right people and information, and for me to understand.

Planning and organization: you can't beat it. Hindsight's always twenty-twenty, and I can see that we made mistakes because we went too fast, and didn't think things through.

Like Sardo says, this place sure beats the cabin. I'm getting better and better at being a solar cowboy.

Now, let me tell you how Sardo earned his spurs. When it comes to cowboys, there's a lot of wannabees, call them dudes if you like, who want to ride horses and look good, but Sardo was here on working days, mending fences, and tending irrigation ditches. End of this last summer, we were trying to get the last heifer in, and she didn't want to go. Now our average weight last year was 735 pounds, so she could be pretty firm about what she wanted. I was pulling on her, my lungs burning, and she had fire in her eyes, but I was getting nowhere. Finally Connie and Sardo rode over, and I told him to put a rope on: "C'mon, ain't you ever seen TV?" I said, when he was having trouble, and he said back, "This ain't TV!" Sardo got off his horse and was putting a rope on her when she charged him, put her nose in his crotch, and threw him five feet up in the air. He got the rope on her somehow, I got the rope through the trailer, and Sardo got back on his horse and we finally got her loaded and down to the sale barn, with three broken ropes still on. Anyhow, I figured, he'd earned his spurs, for sure.

The Preventive Ounce: Regular battery maintenance

In most books about independent power systems, and in this chapter, batteries are described as petty tyrants and the dark underside of the energy self-sufficient household. The truth is that they are safe, familiar, and remarkably well designed. Warnings are necessary because batteries contain a potential for disaster. We are most familiar with them in the context of an automobile, where their contribution to the whole spectrum of vehicular catastrophe is minor, and where their management is usually delegated to professionals.

Batteries demand more attention than any other part of a stand-alone home system. The pioneering school of thought suggests that everyone must crash a set of batteries before they learn. Others, installers with dozens of systems in their portfolios, say that is old news. They say that a well-balanced system, good metering, and decently trained users may skip this expensive learning experience entirely. History shows that batteries installed in thousands of homes and remote buildings present a small and acceptable risk, particularly when precautions are taken to offset the dangers. This caution should be understood as simply part of the independent homesteader's overall commitment to keeping things safe, and continuing the near-perfect safety record of off-the-grid systems. Batteries misbehave only when treated cavalierly or in ignorance, and then they usually fail safely: they just stop working. Some of the threat is due to the fact that batteries are usually consigned to a dungeon, literally out of sight and out of mind. A fire- and acid-proof cabinet reduces the risk below, for example, that of a propane stove, and far below the danger of a woodstove.

Most electrical system components are easy to comprehend, but batteries are quite complex and are poorly understood even by experts. Fortunately, to do a good maintenance job, you need not understand the subtleties, but only the broadest realities. Under normal operating conditions flooded lead batteries, the kind commonly used for domestic storage, convert small amounts of water into hydrogen and oxygen as they charge; if overcharged, they actually boil off water. In either case, electrolyte levels fall over time, and must be replenished. Until a battery set's thirstiness is understood, monthly fluid checks are wise.

Batteries do their job superbly when treated correctly, but since they do this job with dangerous chemicals, they must always be approached with proper care. Before starting, make sure you have plenty of distilled water for watering the batteries and for diluting an electrolyte spill, and baking soda for neutralizing any dribbled electrolyte. Wear goggles or face mask and gloves when working on batteries—you may feel ridiculous, but a splash of battery acid (the electrolyte) in your eye could cause permanent blindness—and wear old clothes, because acid eats fabric. Tools used around batteries must be nonconductive: flashlights with metal cases, long screwdrivers, wrenches, and other implements with exposed metal parts are invitations to disaster, because a battery can unload a dramatic amount of current in a short time, if a conductor comes in contact with both battery posts at once.

Cleanliness is important near batteries. Using a clean rag, dry or lightly moistened with distilled water, clean the tops of the batteries, making sure that any dust and chemical deposits are cleared from the area around the caps before opening them. Impurities in the electrolyte cause plate fatigue and battery failure, so start with a clean battery if possible. You need not worry about being shocked by touching exposed metal parts, because voltages are low (and you are wearing gloves), but it is best to touch only the parts that need to be touched. One at a time, remove the caps from the cells and, using a flashlight, check the electrolyte level. There should be a split ring or fluid indicator showing the ideal level, and you may even be able to see the tops of the battery's plates. The problem here is that batteries are often put in awkward, dark places, and the electrolyte is clear, so it can be hard to gauge the level. Sometimes nudging the battery a bit will make the electrolyte slosh enough that its surface can be seen.

A 6-volt battery usually has three cells, and a 12-volt battery has six, each with its own cap and electrolyte supply, so checking a battery's fluid levels will take time and care. If one cell is full, this does not guarantee that every cell will be full, but it is a fair indication. If one cell in each battery is found to be full, it is safe to assume that all cells are well enough provided with electrolyte that you may check again in a month.

If electrolyte levels are found to be low, using a funnel and a measuring cup, or some other device (like a turkey baster) designed to give good control over the volume of water delivered, gradually add distilled water to every cell in need. I am told by battery technicians that it is better to underfill slightly than to overfill, overflow, and thereby dilute the electrolyte (as well as making a corrosive mess that needs to be cleaned up). If you do make a spill, take great care *not* to get baking soda inside the battery, because it will neutralize the electrolyte. After getting exposed body parts away from the spill, replace the caps and flush the extra fluid from the top of the battery with distilled water. This doesn't make the spill more dangerous; it reduces the electrolyte's acidity and slows its attack. With the batteries sealed off from contamination, baking soda can be applied to the spill to neutralize the sulphuric acid. If any baking soda gets on the top of the batteries, it should be carefully removed.

Unwatered for too long, batteries eventually boil off so much water that they are permanently damaged, and will never again hold a full charge. In practice, battery capacity decreases as electrolyte level drops because a smaller plate surface is available. At a certain point, the charging current and concentrated electrolyte starts to attack the battery, and precipitates a failure that is practically irreversible. Batteries are too expensive; check them frequently.

While checking fluid levels, it is an easy matter to visually inspect and, if necessary, test all electrical connections around the batteries. Because dissimilar metals are enclosed near batteries which exude corrosive vapors and foster metal fatigue, these are the connections in a system that are most likely to fail. Multistranded cables surrender a strand at a time, and more and more current will be forced through the remaining strands until the cable eventually fails. If allowed to go this far, the heat of the current forced through the few remaining intact strands, or the sparking between the cable and its finally severed connector, can damage the battery and in the worst case cause a fire or explosion. Identify a hazard at a very early stage, and replace the offending part with one better guarded against the environmental stresses. In the most demanding environments, where vibration from nearby motors and warm, salt-laden air adds to the battery outgassing, cables are usually soldered to their connectors, then covered with heat-shrink tubing, then sealed with grease or silicon at their terminals. And they still ultimately fail.

Taking Good Care of the Team

Dick Britt's idea of working batteries as if they were a team of horses makes good sense. Each of the batteries, like each of the horses, is an individual, and may charge at a slightly different rate. If a weak battery runs with a strong team, it will gradually lose its ability to pull its share of the load, will weaken the performance of the others, and will finally fail. Periodically overcharging the batteries, giving the weaker battery a stronger, longer charge to catch up, helps take care of these individual differences. This process, which is called equalizing, need not be done more often than once a month. On a well-thought-out system, a battery equalization switch on the charge controller makes this into a simple task.

Battery equalization solves another problem. Over time, the electrolyte in batteries tends to stratify, so the acid is at the bottom and the water is at the top. This decreases storage capacity and also exposes the batteries to the risk of freezing. During equalization, the batteries outgas, and the bubbles stir up and destratify the electrolyte.

Electrolyte maintenance, apart from keeping the batteries watered and keeping stratification under control with periodic equalization, is seldom necessary. Using a hydrometer, a device for measuring the electrolyte's *specific gravity*, which relates directly to the concentration of sulphuric acid, we may determine the state of charge of the batteries. (Again, wear a face guard and gloves, and have water and baking soda close at hand whenever working with batteries. When using a hydrometer, watch for dripping acid.) As the batteries come to full charge, the specific gravity of the electrolyte increases as shown in the chart (see fig. 19). By comparing specific gravity with other indicators of state of charge (voltage and amp-hours consumed) we can refine our sense of the general health of the team of batteries. Very rarely, and usually only because a battery has been over-watered, the electrolyte will measure well below the expected specific gravity, and may require the infusion of additional sulphuric acid, which is a nasty but quite a manageable task. A good battery supplier will be able to provide guidance and the (needless to say, dangerous) chemical needed.

Figure 19.

State of Charge	Electrolyte Specific Gravity
100%	~1.280
80%	~1.250
60%	~1.220
40%	~1.200

(Note: electrolyte specific gravity varies with batteries from different manufacturers. Look to documentation that comes with the batteries for information relevant to your batteries.)

Things That Spin, Wear

Any part of your system that moves will require maintenance. Windmills, pelton wheels, gensets, all have a maintenance regimen which can be ignored with utter confidence that failure will result. Like appliances which operate in the more benign interior environment of the household, each of these devices will come with a maintenance manual, which should be read and heeded. Preventive maintenance will keep failures to a minimum.

For me and for many other men and women who appreciate some level of energy independency, there is an honor and an honesty to the time taken on system maintenance. By concentrating power and delegating responsibility, our society has created innumerable opportunities for pollution. We owe it to our environment to don coveralls and gloves, find the proper tools, and familiarize ourselves with the nuts and bolts and the risks that are part of generating nonpolluting forms of energy. It reveals my ambivalence about the automobile that I will not assume responsibility for its vile underside; I am in no way ambivalent about caring for my home power system, which presents a clean, appropriate, understandable, and rewarding challenge. Its diverse parts work so well, and are so easy to keep running, that I need only remember to give each component the small quantum of attention it requires.

As simple as possible: Dave Katz's Story

David Katz lives off-the-grid on the hillslope above Briceland, California, in southern Humboldt County. He presides over an anarchic band of alternative energy equipment makers and distributors, most of whom live off-the-grid, too. An engineer by training, he is often involved in the discovery, rediscovery, or introduction of appropriate alternative technologies.

I had a Swiss Family Robinson idea: we wanted to see how much we could get for ourselves, instead of from outside. In the early seventies, Bob McKee was following his vision, and turning this into a forty-acre suburbia, selling land for four hundred dollars an acre, no money down if you didn't have it. (Now it's selling, if you can find any, for two thousand an acre.)

My land partner and I paid Bob an extra grand down, and he agreed to do some free cat work, clearing the old skid road down to the spring. Briceland's town water supply used to be a couple of parcels below me, and not too reliable. They'd always wanted to use my spring, but hadn't been able to get access. At the driest time, in fall, there's at least five gallons a minute, more than enough drinking water for me and for gardening. In winter, there's thirty gallons a minute, enough to run a micro-hydro. The sun picks up the slack in the summer. I've got a generator, but I doubt I run it four times a year.

The spring is several hundred feet below us, but there's so much water that two High-lifter pumps can pump enough to the gravity-feed tank up above the houses. We started with two rams. They're very satisfying, the way they pump water using a water hammer, but they don't restart themselves the way the High-lifter does; the High-lifters are altogether more elegant.

My house is pole frame construction. The poles are set in six-foot deep holes dug to bedrock, with cement footings and gravel backfill. It's

Dave Katz surveying his winter vineyard.

very good construction technique for earthquakes. We had the poles set plumb, but by the time we attached the roof, they had shifted and there was nothing we could do, so the whole house is a bit out of square.

We built the house facing magnetic south, which turns out to be sixteen degrees west of south, but what did I know? As you see, I built a rack to reorient the PV array to true south, and it works fine. Part of my life is always doing things wrong the first time. It usually works out good by the end.

It's Humboldt County tradition to tax your property for the value of the buildings, even if they are built without the Building Inspector's blessing. They started a program a few years back to let the Building Inspector write tickets for building infractions, but three thousand people, mostly straight, showed up for the hearings, so the idea was dropped.

I had a business installing oversized alternators and batteries in cars, so you could drive your car around to generate electricity, then park at home and plug your house in. It was a sensible solution, in that automotive lights and devices were all that were available then. My partner Roger and I went to a trade show and saw some kind of panel parked in the back of a booth, and the guy knew enough to tell us it was military, and expensive, and did something with sunlight and electricity. When we finally tracked the product down, we found out they were perfect for remote generation, and that we could buy them by the hundreds. We had our first order for two hundred in very soon.

A while later, a guy in a tie showed up at our workshop, which was a big, barnlike building in Briceland where we'd play ping pong and wait for customers. He was from Photowatt in Arizona, and had been looking for us for two days. They'd somehow found out that some folks in southern Humboldt were buying more modules than anybody else except the military, and he was supposed to find out why. We ended up buying from them, too.

The less electronics, the better, is my theory. Photovoltaics are just so simple. And batteries are so much more complicated than they need to be. In 1986 or '87 I was at the PV Specialist section of the IEEE [Institute of Electrical and Electronic Engineers] in Orlando, and saw some Edison cells, nickel-iron batteries made by Epos, a Hungarian company. The original Edison cells were made in the U.S., and had a lifetime guarantee that they'd replace them if they lost 20 percent of their capacity, but lots of them are still working fine, decades later. Edison's company sold the technology to Exide, I think, in the late fifties, who claimed they couldn't make them, that Edison had held back a secret.

I guess the Hungarians figured out the secret, and so I arranged to have some of the batteries sent for a test. After a year of finagling, I went to Hungary and visited Epos. They told me my batteries were on a lorry waiting for fuel, and sent me to the battery division, which turned out to be an empty warehouse presided over by Kuchar, who speaks a little English. He told me my batteries really were in Brest, just south of where they come from, near Saint Petersburg in Russia. After another year of waiting, we finally got a couple of sets, and are testing them. After getting rid of a bad cell, Humboldt State University is test cycling them two and a half times a day, simulating an accelerated lifetime. They've gone through seven hundred cycles, and they're at better than 90 percent capacity. We're also testing a set of much more expensive FNCs—Fiber-Nickel-Cadmium batteries—that are still at full capacity after twenty-two hundred cycles.

I got the phone number for the battery company in Russia and I've learned to say, "I don't speak Russian, do you speak English?" and I'll try to get an even straighter shot. It's all worthwhile because the batteries are quite cheap and very trouble free. Their only problem is obscure; they're expensive in terms of energy density, compared with conventional batteries—it's the same classic issue as with compact fluorescents. These batteries last much longer and are much less tricky to maintain. You don't need to regulate the charge current, and can run them absolutely dry, and all they do is lose capacity until you re-water them.

I don't make decisions, so things just happen. When I make decisions, I probably do the wrong thing. My business just happened. I'm trying to keep it the size it is, because I'm managing as many people as I can. I am getting better at delegating work, trying to make it more fun and less work for me. I like to go to conferences and hang out with guys in suits. I enjoy travelling, and I like talking on the phone.

We manufacture a lot of the parts we sell, and, except for the PV modules, we could make even more. It's like the Japanese mini-business system, where we have a network of cottage industrialists working for us. We're building a new building, and I intend there to be a Research and Development lab in it. Besides the things we make and our dealer sales, we support a very active solar community with service and installation. That feels very important, that we can help our neighbors get their equipment installed and working right.

My house's main problem is that it isn't finished. It needs more light, a new inverter . . . I'd like to do a radiant floor, and I want hydrogen . . .

Keeping Things under Control

The more complex the electrical system, the more equipment will be required to measure and control the flow of electricity. As in any system composed of dissimilar parts, thought must be given to putting the parts together, maintaining them, and taking them apart. One of the most frustrating experiences of the new independent power producer is encountering a failed component which has been installed in a way to make servicing or replacement impossible. A State of Hawaii rural electrification project in Miloli'i, for instance, installed heavy metal bus bars across the battery fill plugs so the batteries could not be watered without disassembling the

Safety and Balance-of-System Equipment

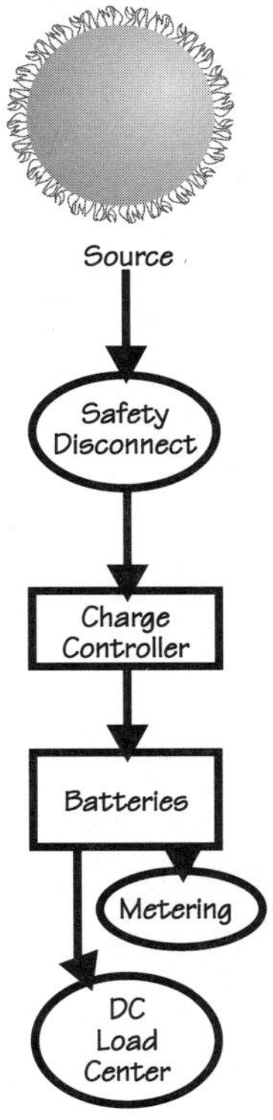

Figure 20.

A home energy system receives its power from a sustainable, local source, and stores the power until it is needed. Besides the source, which harvests energy, and the batteries, which store it, other equipment makes the system safe and keeps it functioning in proper balance.

Early low-voltage systems used simple balancing devices (such as charge controllers), and safety equipment was nonexistent. As more is learned about maintaining systems efficiently, and as more systems are installed, the balancing and safety devices get better. Easy maintenance without risk to humans and equipment dictates most of the safety components. For instance, a safety disconnect allows for opening the circuit between source and storage by simply throwing a switch. The load center (DC or AC) provides a reliable and organized way to connect circuits to the energy source and protect them from themselves, each other, and problems in source and storage. By metering the batteries, we can tell the state of the system and its recent history, and can then plan our usage and maintenance accordingly.

The charge controller monitors the batteries and source, and keeps the batteries as full as possible while ensuring that they are not overcharged. When the batteries are full but the source is still producting electricity, some charge controllers can divert energy to a diversion load, which uses excess power when it is available. Fred Rassman (in chap. 1) diverts extra power to hot water heating elements; I run fans and air-heating elements to redistribute warmth in my house.

battery circuit. Not surprisingly, a majority of the batteries failed quite soon because they were never watered. One way to avoid such absurdities is to uninstall and reinstall components until they are properly situated, but it is even better to install for easy maintenance from the outset. "First in, Last out" (FILO) is a good strategy: install the most robust equipment—wires, disconnect switches, and the like—first, and more failure-prone equipment last.

The simplest form of control is *fusing* and *overcurrent protection.* A battery short-circuited is an impressive thing, capable of creating an amaz-

No Load: The Best Situation of All

If we take energy for granted, we think nothing of leaving a light on so that a room will be "friendlier" when we enter it; appliance engineers have taken the cue, and designed our appliances so they are never really off. Futurist-designers proudly tell us that microprocessors, the little silicon computers that make microwaves and televisions smart, will soon be found in all our appliances. Off-the-gridders ask: Do you really need a clock in your coffeepot? We know that fewer than 20 percent of consumers use their VCRs to time-shift programs by recording them automatically and playing them back when convenient, but every VCR with a clock frivolously guzzles electrons twenty-four hours a day.

What does it feel like to use no electricity? When the power fails, we know. Afterwards, owners of grid-powered houses must scurry around resetting clocks. How many clocks do you have? I have four, two of which are unnecessary; I rely on the clocks which are powered by rechargeable batteries or are plugged into the never-failing 12-volt system. The answering machine is a true wonder of bad design: after losing power, not only has it forgotten what time it is, but it has also forgotten what it was supposed to say to callers! Did its designers live in a place where utility power never fails? It uses 12-volt AC to avoid one 25¢ chip in the clock circuitry, but draws too little current to keep the inverter going by itself, so it must be on-the-grid. Want to buy an answering machine cheap?

ing amount of heat, sparks, and noxious gases in a surprisingly short period of time. Prudence requires fusing between any power generating or storing device and any other device. The next most obvious precautionary measure is disconnect switching: we want to be able to easily and safely sunder the connection between any two devices at any time. Overcurrent protection and disconnects are inexpensive components when the risks of doing without them are considered.

Balance-of-system components are the battery's accomplices, maintaining the system equilibrium by optimizing the charge applied to the

Off-the-gridders breathe more easily when the last load is turned off and their homes quiet down for the night. For these people, electric power is a hard-won energy source to be used only when needed. At night, or in mid-day, when no extra light is required and no work is being done, the only electricity flowing in their homes is from source to storage. They are puzzled by our infatuation with electricity and our eagerness to bath ourselves constantly in its invisible electronic smog. For them, no load is the natural load.

Inverters have become so efficient and robust that conventional electricians often recommend transparent systems, in which the users need not know they are using an unconventional power system. Inefficient appliances such as refrigerators are replaced, incandescent lighting is abandoned, and space conditioning is transferred to another fuel, but little or no change in household usage patterns takes place. For convenience, all loads use house current, and everything including phantom loads is plugged in and drawing current at all times, so that the inverter never shuts down. To accommodate this consumptive arrangement, every part of the system must be oversized. Owners of such muscular systems may persist in using electricity without conscious load management. Yet these same electricians install modestly sized systems in their own homes, and are enthusiastic about the natural rhythms that sustainable power adds to their lives. Rather than pass that enthusiasm along to their customers who are new to energy independence, they find it easier to treat their clients as incurable electron addicts.

I like systems where electrical power is suited as precisely as possible to the most appropriate device for the task. A bright 12-volt compact fluorescent (CF) light, for example, uses less than half the energy the 110-volt version of the same light uses when connected to an inverter. Ironically, the single 110-volt CF is not a big enough load to trigger the load-seeking inverter, and so a further inefficiency is added because something else must be turned on to get the inverter to take the demand seriously.

In a fast-paced consumerist society where we are required to assimilate new technologies every few years, we tend to sacrifice efficiency and ease of use to convenience and ease of learning. While we are learning, this simplicity may be helpful, but when we master the technology we want it to be powerful, not simple. After talking to many independent homesteaders, including many who started in transparent homes and asked their electricians to give them more responsibility, I believe that we are becoming electrical sophisticates. We understand that different tasks require different electricities; we are ready to favor efficiency over convenience. People who harvest their own electricity are proud to understand their electricity system, its demands and special traits. As they learn, they use less power more appropriately.

batteries, and monitoring then controlling sources to provide or store or divert current. In a simple system with a single source, the charge controller handles the whole task; in complex hybrid systems, the job requires a computer with a small raft of sensors and a custom program that accounts for the system's specific interactions. The idea is to choose equipment and manage loads so that the system remains healthy and requires as little intervention as possible.

System metering offers those of us who delight in measuring an opportunity to study the performance of each component with an elaborate metering network. For those who just want the simple facts, an amp-hour meter that keeps a running tab on the energy credits and debits at the battery bank suffices nicely.

Inverters: Different powers

In a homestead power system, the inverter takes energy stored in the batteries as low-voltage direct current (DC), and changes it electronically to alternating current (AC), the kind delivered by the grid, into which we plug standard household appliances. Inverters are the big cousins of those little energy criminals, the transformer/rectifiers, which do their wasteful business in the other direction, changing 110-volt AC into 12-volt DC to charge the electric toothbrush. Battery-based systems employ inverters to convert 12- or 24-volt DC into 110-volt AC. As more homes secede from the grid and the alternative energy marketplace expands, inverters are getting better. More reliable, far more efficient, and much better suited to the real demands of off-the-grid homes than devices only three or four years old, modern inverters do a sloppy job very neatly.

As we already noted, AC house current looks like a sine wave, and most consumer appliances expect their current to look that way. Many inverters synthesize a sine wave using electronic sleight of hand which produces a modified square wave, and some units come closer to matching the right pattern than others. Amplifiers and other electronic devices that work with sound handle the rough edges of modified square waves poorly; their filters are designed to extract the expected sixty-cycle sine wave's hum, but they pass the square wave's buzz right through with the music. True sine-wave inverters are available for circuits that demand purity. Most modern inverters also have the ability to search for a load and to turn themselves on only when one is found; this makes them much more efficient, especially in a no-load situation, when they use only a small intermittent pulse of power to seek loads.

The transmission and distribution system in an independent home looks like conventional house wiring, except that there may be more of it. Safety and practicality guide us: each zone and general function in a house gets its own branch circuit, and so lighting, kitchen outlets, and the shop will each be indepedently wired and connected, through a circuit breaker, to the bus bar's main power line in the distribution center. In homes where two or more kinds of electricity are available, each will have a distribution center and network of circuits. My home has three delivery systems. A 12-volt DC distribution center with branch circuits supplies power for lights and instrumentation, and an AC distribution center controls several conventional 110- and 220-volt branches. In addition, a special uninterruptible system for the computers uses a 110-volt house current charger to float two 12-volt batteries at or near their capacity; the battery power is then inverted back to 110-volt AC for the computers. When utility power fails—which is often here—the computers are blissfully unaffected, and my work is not lost.

Because low-voltage electricity loses potential quickly if it runs through small-gauge wires, low-voltage circuits are usually wired with larger conductors to minimize the loss: a 15-amp 110-volt AC lighting circuit only requires fourteen gauge wire, but to deliver the same amount of 12-volt DC current requires ten gauge wire. As gauge numbers get smaller, the wire gets thicker: fourteen gauge is as thick as a thick pencil lead, .064 inches in diameter, weighing 12.4 pounds per thousand feet; ten gauge, .102 inches in diameter, is more than twice as massive, weighing 31.4 pounds per thousand feet.

Electrical loads are permanently wired into the system as switches and fixtures, or attached more temporarily to the system using outlets. In a household where more than one type of electricity is delivered to outlets, distinguishing between different outlet types is essential. It is best to employ outlet configurations that only accept plugs from devices that are looking for that specific kind of power. Unfortunately, there is no standardization; the commonest 12-volt outlet (a cigarette lighter socket) is not particularly safe; and no relief is in sight. Even sophisticated off-the-gridders report frying the occasional TV set by plugging it into the wrong kind of electricity. Usually only a fuse has blown, but that is embarrassing enough.

Light for life: Robert Sardinsky's story

The original reason for the electric grid was home lighting; at the time, no one imagined the welter of electrical gadgets with which we have since complicated our lives. Many end-of-the-roaders left behind electricity along with civilization in a neo-Ludditic frame of mind, but find themselves drawn back because they miss that sweetest of applications, light.

Many lives have a theme, a realm of knowledge which has intrigued that person for most of her or his life. Robert Sardinsky's theme is light.

Robert Sardinsky enjoying Colorado's high country, as far away from the cares of his business as his horse will take him.

After talking with him, I am more conscious of light and of how I can improve my life by managing it well. Wherever I look, in my own work room, in conventional and innovative homes, off-the-grid and on-, I see that we use light no better than we use electricity itself.

Robert Sardinsky and energy conservation got serious about the same time. After a stint as a pioneer in the northeast he came to Colorado to evangelize, first, for energy consciousness, and now, for light. He and his wife Colleen live in an on-the-grid house with off-the-grid and energy-optimized systems above Snowmass Creek near Aspen.

I figure these four items high on my list of the necessities of civilized life: Light, Music, Refrigeration (for ice cream and beer), and Running Water. Light is the one that captivates me: light is art . . . or it can be. It can make a home cozy, give a store pizazz. I relate to light as an artist and a scientist.

As a way to educate people about energy, light is perfect. We know people are moved by light, and we can swap out a light in a minute, which can be very symbolic. Heating and cooling are important, but there's no subtlety to them: either you're cold or you're hot. Light has so many levels. We're solar-based, and our whole biology is based on light. We are clocks regulated by the sun, the seasons. Light is food, like vitamins.

In fifteen minutes—although I'd prefer an hour or two—I can teach you the basics. It's subtle, but not complicated. We were raised with incandescent light, which has more warmer colors in it and a lower color temperature than daylight, so that's the kind of electric light we are used to and think of as normal. Fluorescents and halogens have a higher color temperature, which means we see them as whiter, bluer. I'm curious: Would I be different if I'd been raised with cool white fluorescents?

It's important to ask, to what use will the light be applied? If you'll be reading or doing black-and-white symbolic work (writing, line drawing, calculating), a cooler color temperature, richer in blue, green, and violet may be preferable, but where true color is important, you'll need warmer light. If I'm lighting a salad bar, I want a source rich in greens, reds, and oranges. We can selectively choose what we want to accentuate. Here's a interesting one: Different races of people look better under different color temperatures, so you can see that cultural issues enter into consideration.

Another important measure of lighting is the Color Rendering Index (CRI). Incandescent light has a CRI of 100 by definition, but standard fluorescents range from 58 to 62. The good stuff we work with is 70 to 90 plus. You always want the best CRI you can get.

People are aware of gross lighting phenomena, like glare or dimness, but are not as sensitive to nuances of light as they are, for example, to scent or air quality. Yet, using light, I can create many effects: soothing, exciting, dramatic, oppressive. The first thing we can do is pay attention to the light where we live and work. Notice a light out of your peripheral vision: Does it flicker? Can you hear a hum? Good quality modern fluorescents don't hum, but the old ones do, and it's very distracting, to say the least, so fix it! Take a look at flesh tones: Do you or your co-workers look like they just got back from holiday—or from sick leave? Flushed, tanned? It's not so much a question of right and wrong, but knowing that you can have it the way you want it.

The worst offender I encounter is glare on video display terminals (VDTs) like computer and television screens. Look closely: do you see multiple images, light fixtures or windows, reflected in your screen? Is the reflected image brighter than the screen image? We're drawn to light like bugs, and our minds must resolve the conflict of multiple images at different focal lengths. When we force ourselves (even unconsciously) to adapt to brightness and focal distractions, we waste our own energy and attention. A big part of my work is simple ergonomics: table heights, screen angles, so that people can come to their work directly, without distraction.

Of course, there's more to light—safety, security, and technical issues like efficiency, energy and maintenance costs, and the interaction in a building between lighting and heating, ventilation and air conditioning (HVAC). There's so much to know and to do, there's no way one person, or one company, could do it all.

Thinking about how I got to this place, I remember my first business. I had a travelling light show when I was thirteen, fourteen, fifteen. It was before I could drive, so my parents had to pick me and my strobe lights up at three in the morning from high school dances. So you could say I've been studying light all my life.

I grew up in suburban Philadelphia, and another important source for me was Fairmount Park, which gave me my first sense of real wilderness. The Institute for Social Ecology at Goddard opened up my world; I studied with some of the most radical and innovative thinkers of our age. And I kept running across real inventors making extraordinary things: a guy in Vermont doing passive refrigeration by making a giant ice cube in a kiddy pool nestled in a silo; another guy working on a passive hydronic cooling system in his kitchen's northerly outside wall. I heard about the New Alchemy Institute and had to visit. My first impressions: a sailwind windmill's slow bright colors creaking in the wind, lightning bugs and black-light bug zappers, fish grabbing at the comfrey or whatever in the pools. It was awesome. I'd been at Penn State for two years, studying Art and Environmental Education, and I was ready to really do something.

At New Alchemy I helped oversee the Ark, but I was involved in everything. I studied CO_2 dynamics in bioshelters, and the process for cycling municipal waste into CO_2 and food, and the relationship between light and microclimate . . . I created the Institute's education program and self-guided tour. I lived in a yurt, built a solar shower, then a solar shack with a half-acre garden with raspberries just beside the door so you could step out to pee and pick berries at the same time. There were volunteer tomatoes, and windrows of carrots, parsnips, kale, and parsley that wintered over under the oak leaves.

The bioshelter is a special place: it brings the summer garden indoors. In winter in that climate the bioshelter is in the doldrums two months of the year, but the rest of the time . . . the flowers! the smells! I remember particularly skiing in to the bioshelter and picking fresh tomatoes on Christmas day, and watching a butterfly looking out of the ark at the foot-and-a-half deep snow. The hardware is fun, but it's the living stuff that turns me on, and the integration of the two is best of all.

Somewhere along in here I invented the sunbrero (a solar-powered beanie) and started wheeling and dealing PV panels because Boston was the center of much PV development at the time. I found this warehouse full of PV seconds, an eclectic mixture of stuff, and I started buying station-wagon loads way below the going rates and distributing them around the country.

My idea about houses is, small is beautiful. The ultimate challenge for architecture is to make small spaces feel roomy, to have both intimacy and openness. Home can represent a model of what needs to happen on a larger scale. The wholeness, sense of place , and richness can be, should be, inspiring. Not everyone wants a garden, but I've never met anyone who isn't moved by a garden. My grand vision integrates people, plants, and architecture to create a stronger sense of *oikos,* home, a sense of family diverseness, richness, light, growth, and the senses of space and place. I get disoriented easily in cities: it all starts blurring, it's scary. Where there should be mountains, there are buildings, and there's no sun, it's usually overcast. I think we all sense this, and are drawn to the rural, want to move there. But we bring our consumptive garbage with us. We need to redefine and rethink . . . and we've got to do it quick.

I figured, we need to make models, and start close to home: our own homes. Do we know where our power comes from? We sure know bigtime when it isn't there! I found this rental house, and started making it into a demonstration of energy efficiency, doing the best I could with what I had. I didn't plan to secede from the grid, because the house shell itself is too inefficient, insulation at R-30 or R-40 at best, heat-losing glass. Naturally, the first thing I did was install all high-efficiency lighting, mostly compact fluorescent. I put up photovoltaic modules, a battery bank and inverter, and separated circuits out logically so that solar powers all but the resistive loads. It's fun to turn the lights on, and know they come directly from the sun. When the sun runs low, the system switches over to draw from the utility.

All the glass in this house works well—too well—on sunny days, but the house has too much volume and very little thermal mass, so even if we are very frugal, during the worst months space heating can cost $140. Conceptually, it would be good to increase the thermal mass, but that's not easily done in a rental. We did a blower test, and found serious infiltration, which we mostly solved with forty-eight tubes of caulk. We converted the fireplace to a state-of-the-art woodstove (a Vermont Castings Defiant Encore) and switched from electric to propane cooking. We made these window covers by cutting up $32 queen-sized indian blankets, copper pipe, and left-over leather grommets from my sunbrero-making days, and they work great. I priced them, and the drapery people wanted $7 a square foot for something that wouldn't have worked as well. Now, we're puzzling on how to get the heat trapped against the high ceiling to circulate through this big, impressive, useless volume.

I've got a mess of micro-PV systems around the house for the fun of it: step lights that look like runway lights by the front door, and reading lights and a stereo in the bedroom. My truck has PVs in every window to recharge the cordless tools.

Knowing this is a rental, and that whatever gets installed might have to get uninstalled, I've been looking for low-tech solutions. I've got a qualitative problem with discount stores—our values are getting royally screwed up. Colleen and I like really fine things, that last, that don't lose their beauty. There are wonderful old classic cars, for example, that look as fresh as they did when they were new. It's dangerous when we make do with temporary, cheap goods that don't last.

CHAPTER 5

THE TREE GAME: *Reckoning Our Footprints on the Planet*

Energy politics has become a realm of beliefs altogether removed from rationality. Making our policies more rational has been the object of incessant struggle in almost every state, as alternative energy advocates have done battle with entrenched, interdependent, and monopolistic energy corporations. Those with vested interests in the present system are reflexively fearful of any kind of change, and are quite rightly afraid that rational energy pricing will change the way we perceive them—that they will lose their subsidies and be required to do business responsibly. The path to a rational energy policy, where we pay the true, whole-life cost of every kilowatt, therm, and gallon, will be fraught with potholes and landmines. We would expect the global energy establishment to resist and conceal, but bureaucratic hypocrisies and the multiple hidden costs of energy compound the difficulty of honestly and thoroughly calculating the price of renewable energy.

Most of us who have committed ourselves to using renewable energy options do so, therefore, as a matter of faith. We believe that by conscientiously reducing our usage, and then providing for our remaining requirements with locally available, renewable sources, we can work some magic on the global energy picture . . . if enough people elect to work beside us.

In the previous chapter I asked you to calculate your electrical requirements. In this chapter, I would like you to evaluate all your energy uses and think about how you fit into the larger, planetary energy budget. Since this can be a complicated topic, I suggest we explore it by playing the Tree Game, in which we represent in trees lost or gained the energy consumption or energy conservation in our homes.

The object of the game is to look at hidden and unexpected energy costs and benefits in the context of ordinary, day-to-day energy usage, in order to see how well you are doing. This simplified version of a more detailed assessment is meant to help you think about your use of energy over the long term. Hence, our point-counters are metaphorical trees, from the diminutive piñon of the high desert to Yosemite's mighty Sequoia giganteas. When done, you will have a number that translates very roughly as the number of trees your household has added or subtracted to the planet.

Although the game is meant merely to suggest interrelationships, and has no quantitative basis, ideally this process will not be too difficult or depressing, and will bear some relation to fact.

Several rules: When asked a question, take your first quick estimate. When an answer involves calculations, always round up. If you genuinely do not or cannot know the answers, because the house you live in was built long ago or by someone far away, guess. If you cannot guess intelligently, enter zero.

THE TREE GAME BALANCE SHEET			FACTOR	TREES
I	**Home materials and preparation**		(10)	
II	**Home performance**		(20)	
III	**Home perseverance**		(5)	
IV	**Work: energy issues**		(30)	
V	**Work: caretaking**		(15)	
VI	**Play**		(10)	
VII	**Plan**		(10)	
			Total	

Let us break home activities into seven categories. Three have to do with the building itself, and correspond to its creation, its occupation, and what happens to it after you leave. Three have to do with how you spend your time while living with and in the house. One last category gives you a chance to garner points for planning on mending your ways within the next decade.

This game is about energy impact, using our widest possible understanding of the ways energy enters into our homes. For example, a board in a house has energy embodied in it, first by the tree that grew it; then by the lumbermen who felled, transported, and milled it, next by the distribution and retailing arrangement that gets it to the home site; and finally by the effort required to cut it to length and affix it to the house. The first part of the game is like bidding in bridge, where you make up some numbers which may or may not be based on certainty and knowledge. The numbers I ask you to make up now will represent the relative importance to you of each of the seven categories. Given 100 percent to distribute among the seven, please give to each a share based on your sense of the total amount of energy it will require over the whole time you reside in your home. In the interests of simplicity, may I suggest nothing less than 5 percent for any category? (My own factors are shown in parentheses.)

There are some basic assumptions here. I assume that a home's ongoing energy performance will be about twice as costly as its construction, and that dismantling the house will take half the energy required to build it. Over a generation of occupancy, perhaps twenty-five years, I assume it will cost half as much to maintain a house as it cost to build it. These ratios are all very approximate, and the real numbers are nearly unknowable. Without taking too much time to worry over such confounds, please begin by filling in the factor column.

Next, we must break out aspects of each of the seven categories with enough accuracy to be meaningful but without agonizing over numbers. Hunches are allowed. Remember that energy is the issue, and that plus points are environmental gains while minus points are assessed for negative impacts.

The Home

Section I covers activities that take place before the house is built: clearing the land, roadbuilding, and preparation of the site for the foundation. The first line, Deforestation and Reforestation, is simplicity itself: How many trees were cut, and how many were replanted, to prepare your land? If trees were cut on your land, then milled and used to build the home, credit will be awarded in the next section.

Assess a minus tree for every thousand square feet of open space lost to

I.	HOME MATERIALS AND PREPARATION		
A.	**Site Selection and Preparation**		
1	Deforestation and reforestation		
2	Exported/imported materials		
3	Energy source: incident pollution		
		I.A Site selection & prep subtotal	

B.	**Structure**		
1	Indigenous/strategic materials		
2	Embodied energy		
3	Distance to building supply store		
4	Average material travel		
5	Energy for tools		
		I.B. Structure subtotal	

C.	**Furnishings and Appliances**		
1	Embodied energy		
2	Average material travel		
		I.C Furnishings and appliances subtotal	
		I. total	

a new road or building. The next line, Exported/Imported materials, is more complicated: For each truckload of material delivered or removed, one minus point. This includes rock for the road, unless it was acquired as part of road construction. For every truckload of organic material burned or left to rot, minus two trees. The last line here reports the energy sources and pollution caused or avoided during site preparation: one minus tree for every fifty hours of cat, grader, or backhoe work, and one minus tree for every hundred hours of generator or chain saw work; if any work is accomplished using a renewable source, including the sweat of human or animal brows, award a tree for every two hundred hours.

Add up the trees, and reassemble at the next section.

Section I.B is about building the house. The first line (I.B.1, Indigenous/Strategic materials) is where you can take credit for materials from the land itself or from very near at hand. For every local tree cut and milled, for every ton of indigenous rock or dirt used in walls and foundation, a plus point. Assess a minus tree for every thousand dollars worth of strategic or endangered materials, copper, teak, and other rain-forest woods not grown in plantations, or petroleum by-products such as composite roofing. The number on the next line, Embodied energy, is impossible to calculate with certainty because the answer is buried. How much does the energy embodied in aluminum windows cost? Aluminum is exceedingly high in embodied energy, because refining and machining it is energy-intensive; glass, especially high-tech glass with its rare fillers, coatings, and sealants, is likewise high in embodied energy and therefore hard to appraise. Our entry here should attempt to include all materials, from the steel reinforcement-rod and the cement in the foundation through the finishing materials, including wood trim, paints and stains, and all the other materials that go into the house. Assess a minus tree for each thousand dollars worth. (Trading and bartering counts.)

Distance to the building supply store is a sore point. When building, we drive back and forth a great deal for supplies. From the end of the road, this may be quite a distance, and only if you can convince yourself you were very organized and seldom took extra trips should you assess yourself less than a minus tree for each mile to the store (one way), with a maximum of minus twenty-five trees. Average material travel (I.B.4) is another guess: imagine a circle within which half the raw materials for your house originated, and assess a minus tree for each hundred miles. Tool energy (I.B.5): one positive tree per tool for each two hundred hours of renewable energy, and one negative tree per tool for each two hundred hours of grid-powered energy, or for each hundred hours of chain-sawing or generator-running.

Please total up section I.B.

Now let's consider the furnishings and appliances that go into the finished structure. Using the same measures as before, assess points, mostly minus, for what goes into the house. Minus one tree for every thousand dollars of value. Handmade, homegrown appliances or furnishings, like a thermal refrigerator or willow-work chairs, earn a plus tree for each thousand dollars of value they replace. Commercial appliances will generally show their origin on their nameplates; again, imagine a circle from inside of which half your appliances came, and subtract a tree for each five hundred miles in the radius. Remember, round up for any fractional results.

Please add up the three parts of section one, and transfer them back to the balance sheet on page 100.

On to section II . . .

How well a house performs depends on variables of climate, microclimate, and any other number of imponderables, but forge on with me in search of trends and generalities. For each hundred square feet of windows, one minus point. For each hundred square feet of windows where drapes are customarily drawn in warm or cold weather, award one positive point. One tree on the positive side for every hundred square feet of double glazing, and three plus trees for each hundred square feet of high-tech, low-emissivity glass. For insulation, give yourself as many positive trees as your lowest R-value. ("But how should I know my apartment's R-value?" asks an apartment dweller who is expecting to make the end-of-the-roaders look like energy hogs. Ignorance, on an issue like that, is no excuse. Either your apartment is well-insulated or you heat the outdoors at great expense; even if the landlord pays utilities, it is your wasted energy. I suggest making a deal with such a landlord: Ask for a rebate of half of every dollar saved over the previous year's energy bills, which gives both of you a clear incentive to do better for the planet.) As for home owners, if you do not know your R-value, it is much too low. Enter zero.

The next line gets calculated based on a year's heating and cooling costs. For each thousand kilowatt-hours of renewable, self-generated electricity used for heating and cooling, award one point. For every cord of firewood, every ten thousand kilowatt-hours of electricity, each thousand therms of natural gas, each hundred gallons of diesel or propane, assess one minus point. If you are not able to distinguish between electricity used for lighting and for heating, add the penalty for all your electricity use here.

Total up the points for section II.A.

In the next burst, we will finish off home performance. Domestic hot water heating, as practiced in the United States, is a scandal. Penalty: one minus tree for every ten gallons of hot water languishing in a hot water heater that consumes any form of fuel. If you have an electric hot water heater, subtract two more trees; subtract one tree for each gas heater, either instantaneous (demand) or with a tank. Subtract one tree for each ten feet the hot water must travel from tank to bathroom and kitchen sink. If you have solar hot water or solar-assisted hot water, cut your hot water penalty in half. If you heat water with wood from your own woodlot, you get away for free.

Next, for line II.C, Refrigeration, minus five trees for each refrigerator over ten years old, minus three trees for each so-called efficient refrigerator, and minus one tree for each super-efficient unit. Lighting: minus one tree for every ten incandescent bulbs in the house. Remember to round up. Plug loading: minus half a tree for every phantom load. A phantom load is one that is still on when it is supposed to be off—instant-on televi-

II.	HOME PERFORMANCE		
A.	**Heating and Cooling**		
1	Windows		
2	Insulation		
3	Energy for heating and cooling		
		II. A. Heating and Cooling subtotal	

B.	**Hot Water**		
C.	**Refrigeration**		
D.	**Lighting**		
E.	**Plug Loading**		
F.	**Energy Source**		
G.	**Garbage**		
		II. total	

III.	HOME PERSEVERANCE		
A.	**Longevity**		
B.	**Recyclability**		
		III. total	

sions, every device with a transformer on the outlet side of its switch, plug-in clocks, and microwaves and VCRs with clocks; do not forget answering machines and telephones that require electricity. Half a tree down for every monthly hour of use for each energy hog device: blow dryers, irons, toaster ovens. Add a tree for every plug strip or switched outlet actually used to turn phantom loads off when not in use.

On a scale from minus ten to plus ten, rate your energy generation capacity. If you get all your electricity from the grid, minus ten; if you generate all your own, plus ten; fifty-fifty gets a zero.

Finally, evaluate your garbage management: one plus tree for each separation you perform (glass, aluminum, compost, plastic, and so forth; refer, if necessary, to the Garbage Pie in chap. 12). One minus tree for each garbage can full of unseparated garbage sent to the dump in an average month.

Add up the points for section II and record them in the balance sheet.

What will happen to your home when you leave it? The market for commodity homes is well established, but their lifespan is not well known. The average for a house built since 1945 varies regionally, but looks like it may be as short as fifty years; there is reason to believe that houses built since 1970 will be even less durable. Analyzing longevity is a very tricky proposition because makeovers, in which most of the structure is replaced, are increasingly common. The housing industry is very reluctant to publish statistics, and so information is at best anecdotal. Will non-commodity, independent homes fare as well, worse, or better? The average tract home occupancy is about ten years, but a customized home, built for the occupants, is likely to be occupied at least twice that long , but twenty-five years is a long time for any family to inhabit a modern home. The housing stock still standing and in use after a century is variously estimated to represent from 30 to 50 percent of that originally built. Let us, in the spirit of the chapter, accept a century as the life of a house.

How long will your home last without heroic reconstruction? Assess a penalty of a tree for each decade less than a century, or award a tree for each decade more, not exceeding ten points either way. After a century, what percentage of your home will be recyclable? Take a plus tree for each 10 percent of your estimate.

Please add up your points and transfer them to the balance sheet.

Work and Play

As a culture, we use an impressive amount of energy getting back and forth between home and work. Families who live off-the-grid and still commute to urban work can easily give back all their energy gains by commuting.

Commute miles per month means vehicle-miles driven. If you manage to carpool, you need only count the miles you actually drive. Miles driven to shop and school count: school is the work that children do, and shopping is a requisite of American life. For each thousand miles, or for each ten thousand miles flown in the line of work, the penalty is one tree on the negative side. Miles driven in an electric vehicle count for a quarter of a dinosaur guzzler's miles. If your vehicle gets more than forty miles to the gallon during commutes, reduce the penalty by 25 percent. Here is a tough one: How many miles a month are driven by delivery trucks to serve just you? This should be calculated as the average number of deliveries times twice the distance from the nearest likely source of a delivery (your next-door neighbor) to your door; charge a minus tree for each hundred miles. Finally, charge a minus tree for every ten gallons of gas used in an average month. (Double jeopardy! But hopefully we all agree that fueling vehicles with fossil fuels is crazy.)

The line called Work/Environment is quite subjective. On a scale from minus ten to plus ten, evaluate the work you and your family do from an environmental standpoint. Children in school get a plus tree each. Work that actively involves helping others live ecologically responsible lives is worth plus three. Work that indirectly changes things for the better, like teaching in a school, is worth one plus point. Work that is consumptive but attempts to be efficient, like homebuilding, is worth minus one, whereas work that heedlessly consumes or despoils should be worth minus three. Count up (or down) the points for every family member (maximum: ten trees plus or minus).

Please add up your section IV points and copy them to the balance sheet on page 100.

IV.	WORK: ENERGY ISSUES		
A.	**Commuting/Homework**		
1	Commute miles per month		
2	Shopping miles per month		
3	Delivery miles per month		
4	Gallons of gas per month		
B.	Work/environment		
		IV. total	

V.	WORK: CARETAKING		
A.	**Stewardship**		
B.	**Children**		
C.	**Food**		
1	Gardening		
2	Local produce		
3	Supermarket		
4	Bulk buying		
D.	**Pets and Animals**		
		V. total	

At last, here are some chances to add lots of trees. Stewardship has to do with the time we spend maintaining or improving land—ours, or anyone's. (None of it really belongs to us anyway.) For every ten hours spent in a month improving the land, credit a point. Number of children is a delicate issue, but it seems evident that exact replacement is the mood of America: that means two natural children. For each natural child more than two, assess one minus point; for each child less, award one plus point. Adopted and foster children are weights lifted from the body politic, but the best we can do in an energy calculation is not count them at all.

For food, allocate three trees for each person in the family, as follows: Determine the number of points to be distributed (three times the number of family members) and then do your best to divide them by volume of groceries from each of the four sources. Now, double the number in Gardening, and put a minus sign in front of the supermarket entry.

Award a tree for each working animal on your homestead (cats, even when standing in for architects, are not considered to be working. Whether architects should be classified as non-working pets is still being hotly debated. See chap. 7). Assess a tree for each useless pet, no matter how lovable. Award a tree for each stuffed animal fulfilling the functions of a pet, but only if it is made from non-synthetic materials. The sum for this line should not exceed plus or minus five.

Please sum your points for section V and copy them to the balance sheet on page 100.

The home stretch! Travel: for each thousand miles you drive for fun each year, and for each ten thousand miles you fly for vacations, assess a penalty of one minus tree. For each ten hours of community service along the trip, award one plus tree. Avocations and Hobbies: for every five gallons of fossil fuel consumed in the pursuit of fun by some form of recreational vehicle like a jet-ski or off-road vehicle, one minus tree. For every tree planted, one plus tree. Neither part of section VI should exceed ten trees, plus or minus.

Planning

Section VII is the most subjective. I mean it to cover plans which will be realized in the future. On a scale from minus ten to plus ten, where zero represents no change, award or penalize yourself for the energy impact of ways in which you expect to change your life in the next decade. For example, I awarded myself three positive homestead improvement points for planning to change all single pane windows in my house to high-tech glass and for vowing to plant five trees a year. Committing yourself to use an electric vehicle for shopping trips or to reduce commute days from four to

VI.	PLAY		
A.	Travel		
B.	Avocations & Hobbies		
		VI. total	
VII.	PLAN		
A.	Homestead Improvements		
B.	Vocational Improvements		
		VII. total	

two per week might count for plus three vocational improvement points. You may claim only ten trees for each category, either plus or minus.

Please copy tree sums for sections VI and VII to the balance sheet.

Tree Balance Sheet: Final Tally

You should now have seven sums in the trees column of the Balance Sheet. In each category, multiply trees times the factor assigned at the start of the game, and enter the result under sum. Then total the sums.

This grand total is a meaningless number that is related in some intuitive way with the weight of a family's footsteps on the planet. If you attain the desired goal of a positive total, yours is a seldom productive and welcome clan on this beleaguered planet, and you have much to teach the rest of us. If, like me, your numbers come out on the negative side, especially if your total is in the hundreds, you are performing according to the expected cultural norm, and can congratulate yourself on being a world-class consumer.

Since most of us wish to leave something good for our children, we have work to do. I was horrified to find that my family got –313 trees. We travel too much, both back and forth to the store and for work and play (the wages, I suppose, of including two college-age daughters in the family instead of making them account for themselves). Nevertheless, our household ended up in the middle of the pack of interviewees who played this silly game and afterward were willing to confess their results. Not one score of better than positive 100 was reported, and scores of negative 1300 and negative 1150 were grudgingly admitted.

I will do better. Will you join me ?

CHAPTER 6

CHOOSING A SITE FOR ENERGY

AN INDEPENDENT HOME IS BUILT LIKE ANY OTHER, WITH SOME ADDITIONAL considerations. The builders of an independent home must understand the relationships between the home and its inhabitants, its surroundings, the materials it is made of, and its energy sources. Desire for self-sufficiency presents us with challenges and opportunities when we start to plan an independent home; we have a chance to become partners with the land, and let the landscape itself show us where it would be best to build. While conventional residential construction bruises the land and flattens its particularities, independent homebuilding waits for signs that show the way, and follow the lead of the land.

Until the early 1980s, the pattern of home life in the United States was absolutely determined by two systems laid down upon the land by humans: a pervasive network of roads, and the only slightly less far-flung network of the utility grid. Our comforts are so intertwined with electrical power that until the 1980s the presence or absence of the grid nearly always determined where and how we built our shelters. In the United States in 1980, there were about one hundred million houses; fewer than ten thousand, or .01 percent, mostly low-income families and remote troglodyte cabins, were not connected to power lines. In a little more than a decade, that number has increased by a factor of ten, and although still represents a mere .1 percent of American homes, the tide of settlement has turned away from the deterministic grid. More and more people will gladly disconnect from the life support, as Dennis Weaver so aptly describes it, and live independently.

Voluntary simplicity: Lennie Kaumzha's story

Before deciding to leave the grid's peculiar pleasures behind, we should consider how many of these, or how few, we really need in order to be happy. Could we be happy without all the newfangled electrical inventions and distractions of twentieth century? If not, which would we care to keep? Voluntary simplicity is an idea that has ebbed and flowed periodically throughout history; it came to a particularly high mark at the same time many were moving back to the land, in the late 1960s and 1970s. For some of us, electing a simple life provided new perspectives and a foundation on which eventually to elaborate a more complex but still suitable life, while for others, simplicity continues to be the basis for happiness and productivity. Lennie Kaumzha lives simply in a cabin near Brattleboro, Vermont, and tells the story of purposeful simplicity.

Lennie Kaumzha's cabin in Happy Valley.

I have been a member of the Common Ground Restaurant, a worker-owned cooperative, for fifteen years. We emphasize vegetarian meals, and recycle many of our "waste products." I manage shifts, wait tables, and have numerous behind-the-scenes responsibilities, but my favorite job is dishwashing.

I live in Happy Valley, an enclave of eight cabins all basically without modern conveniences, one of which I rent. I supply electricity for a couple of lamps and a boom box from 12-volt marine batteries which I charge at the restaurant in the winter. I guess you could say I come to town for electricity. In warm weather, I plug into my car battery. I have running water, a gravity-fed system, most of the year; the pipes froze a few days ago, and so I haul water from the restaurant, maybe two or three months out of the year. I have what I consider a very good, simple shower system, right next to the woodstove, using a five-gallon jug mounted on the ceiling rafters, which drips into a plastic pan surrounded by sheets. I mix water heated on the woodstove with cold to the right temperature in the jug before hoisting it up. I spend about one hundred forty dollars a year on wood scraps, bought from a local lumberyard for heating the cabin. Refrigeration is a plastic ice chest; I freeze ice for it outside in winter, and at the restaurant in warmer weather.

Happy Valley sits nestled up against a mountain and reminds me of a hamlet out of time (except for the vehicles sitting in front). I write poetry, haiku for the most part, play fiddle, Irish for the most part, paint with acrylics, and make tin-can candle holders and lampshades with an oxyacetylene torch. I also garden and make maple syrup.

I am forty-seven years old, never married, no kids, and live currently by myself in the cabin. I don't know if I will ever have enough money to buy land and a cabin elsewhere, but I would be content to stay where I am now for the rest of my life. I live fairly simply, not because it's the politically correct thing to do, but because I don't feel right any other way. I consider myself a peasant and feel uneasy around professional circumstances. I have been influenced by my Peace Corps stay in Gambia, West Africa, by the Native American spirit, and by Zen philosophy.

I feel that the more we live simply, using the least amount of money possible, working at jobs as little as possible (I average twenty-four hours a week at the restaurant; someone told me work is what we do in our spare time), the more we live closer to dirt and nature in general, the happier we will be.

Free At Last

Most of us find it impossible to imagine a house without electricity. Now that we can generate the electricity we need, the grid no longer defines the perimeters of settlement. We can select our homesite for its merits alone, sure in the knowledge that we can harvest electricity anywhere we choose to settle.

Electricity can be relied upon to cost more and more as easy energy gets scarcer, and as visible and invisible subsidies (called externalities) are rationalized and monetized. Prices asked by the utilities for line extensions

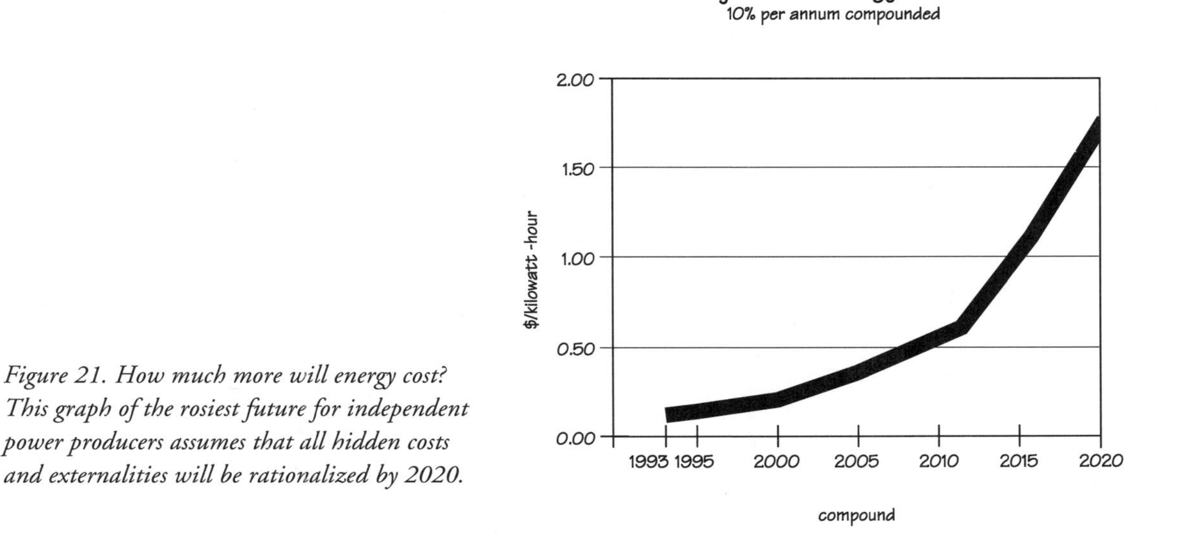

Figure 21. How much more will energy cost? This graph of the rosiest future for independent power producers assumes that all hidden costs and externalities will be rationalized by 2020.

have also risen abruptly, so the cost of stringing power lines across even a mere half-mile of emptiness between a homesite and the nearest grid-line makes a convincing argument for independence. As you will see in subsequent chapters, there are other reasons for disconnecting from the grid, or for connecting as lightly as possible, but cost continues to be a most compelling concern.

If we have the inclination or the need to build independence into a home, we should pause first and review a list of the possible independences we may wish to attain. Homesteaders who have built independently told me with admirable vigor and intelligence about the freedoms they cherish. As I visited more and more homes, the list of freedoms kept growing. Here are some basic freedoms worth considering, organized into the three phases of home-making: building, occupying, and abating.

In the process of building it is quite possible to waste so much energy and material while constructing shelter that any later savings will be no more than poor restitution. A truly independent home will be sensitive to the environment from the very first, and will built in the spirit of the following freedoms:

Freedom from the insults of local micro-climate. Energy-sensitive siting allows us to take best advantage of all the potential of a homesite. The north slope (in the northern hemisphere) is dank in winter. Any exposure to direct sun without access to the tempering breezes can make summer hellish in hot climates.

Freedom from the hypocrisy of making war on the land before settling. We need not destroy the landscape and its lifeforms in order to occupy it.

Freedom from reliance on rare, endangered, or strategic building materials. Local or recycled building materials are cheaper and better suited to local conditions.

Freedom from pollutants and extractive or exploitative materials and techniques. Homes made this way are healthier for their inhabitants, and demonstrate that we can live happily without hurting others.

When we live independently, we dedicate ourselves to living gently on the land, knowing that such a life will give us freedom from many old patterns and dependencies:

Freedom from dependencies on governments, far-flung monopolies, and extractive and polluting technologies. Local energy sources are in tune with their environment; to the extent that these are truly free even from hidden pollution and deferred costs, they bring a gratifying economy to our lives.

Freedom from the necessity of daily travel. By integrating life and work, we may find that both are enhanced. Few who manage to work at home are sad to give up the daily commute, the inhaled exhaust, and the frustrations of traffic and parking. If we limit travel to necessities and adopt vehicles that minimally impact the biosphere, we may be able to preserve our treasured cultural restlessness.

Freedom from estrangement with Nature. We know the restorative and inspirational value of taking nature into our lives. Why not live always with beauty?

Freedom from food and clothing dependencies. Living independently, we may find that our tastes in clothing and food change dramatically, and our needs become more manageable within the simpler context of our lives. To the extent that we can produce our food and clothing on the homestead, we are liberated from reliance on external banking and marketing systems, which devour time and reduce our ability to concentrate. Those who devote their lives to providing basic necessities praise that path highly. Personally, I prefer to focus on what I do best, and let others, master seamstresses and farmers, do what they do best on my behalf.

Freedom from propaganda and regimentation. Only by taking responsibility for our own entertainment and education, by creating local societies and schools, can we free our minds.

Freedom from our own garbage. By intensively reducing, reusing, and recycling our waste stream, we bring consumption under control. This, more than any other action, runs counter to the prevailing culture, and asserts our willingness to depend on ourselves and the place we have chosen as our own. This is also an area where our culture will assuredly soon come to meet us. Overflowing landfills and a new understanding about pollution in our air, oceans, and rivers is changing our national garbage behavior rapidly.

We may wish to build for the ages, but this is unlikely: all things fall, and are built again. We must accept the inevitability of change, and learn to plan for abatement from the start. Planning for the long term gives us access to further freedoms:

Freedom from wastefulness. If we have built with materials and techniques selected to be durable, maintainable, and finally, capable of being dismantled and salvaged with very little loss, the materials in a home are truly invested in it rather than consumed by it. (We hope that the strictures against the use of recycled building materials, conceived before scarcity became the rule and presently enforced by building authorities, will have yielded to reason by the time we are ready to demolish our home.)

Freedom from knowing we will be remembered for our selfishness. By treading lightly on the land while building and living, we leave a haven that may be enjoyed in perpetuity or restored to its natural beauty after we are gone. We owe our children no less, and any action that outlives us bequeaths its own burden of care. Most of us are not willing, for example, to leave a monument to our greed and heedlessness in the form of decaying plutonium which will be deadly for 225 centuries.

Is Independence Affordable?

The building blocks of an independent house are assembled first in our minds, and then with our hands. To the extent that we anticipate what we must do and what obstacles will stand in the way, our progress will be smooth, and so time spent in careful selection and planning repays itself abundantly. When we rush to completion before visualizing the process step by step through to the end, we condemn ourselves to live within a monument to our mistakes and impatience.

Peasant wisdom decrees that there are three important considerations to make before we even begin to think about creating shelter for ourselves: location, location, location. Choice of a site is everything. The architecture of dominion arrogantly pretends that a seemly house can be built anywhere: in the darkest gulch, on the polar ice cap, at the edges of the

	Freedom, quiet and privacy
	Breathable air
	Good water
	Attractive surroundings and suitable interior space
	Utilities: electricity, phone, cable TV, sewage
	Good neighborhood, good neighbors
	Close to school and shopping
	Available transportation
	Community services: police, fire, ambulance, maintained roads
	Retains or appreciates in value

Figure 22. Checklist of site requirements

solar system. Perhaps such a house can be built anywhere, but it does not become a home until it is lived in graciously. Centuries of experience suggest that a home can be no better than its site, nor can it be better than its accommodation to that site. A checklist of site requirements might look like that you see in figure 22.

In a matter as important as selecting a home site, we must make certain that all elements that bear on our ability to live happily in a home have been identified and considered. Site selection is never simple. The perfect site is seldom immediately discovered, and so we must hope that the exercise of weighting a variety of elements and evaluating a variety of locations will lend a degree of logic and objectivity to the process. What do we require of a home? Enumerate the important elements that will figure in the life of your home, being sure that anyone who will have to live with the decision has a chance to contribute to the list. Some requirements are pass-or-fail—someone with asthma should never be made to live in a pasture of goldenrod—while others are preferences. Each element should be given a priority or a weight. Additionally, an element may turn out upon analysis to be a problem that goes away when money is thrown at it. (Water in the desert comes quickly to mind.) If the amount of money required is beyond the sum available, or overshadows the rest of the project, then a site must be disqualified. With this realization we see that any site is a compromise with respect to its many elements. The object of this exercise is to enumerate completely and prioritize fairly so that the best compromise emerges.

The evaluation of a pair of sites might look like the comparison in figure 23.

SITE COMPARISON MATRIX							
		Site 1: Rural			Site 2: Urban		
	Weight	*Points*	*Total*	*Cost*	*Points*	*Total*	*Cost*
1. Freedom, quiet and privacy	25	9	225		3	75	5,000
2. Attractive/suitable	20	10	200	100,000	6	120	75,000
3. Breathable air	15	10	150		4	60	
4. Good water	15	8	120	5,000	4	60	
5. Good neighborhood, neighbors	10	5	50		8	80	
6. Available transportation	5	0	0		10	50	
7. Close to schools and shopping	5	0	0	26,496	9	45	3,900
8. Retains or appreciates in value	3	7	21		6	18	
9. Utilities	2	0	0	20,000	10	20	
10. Community services	0	0	0	15,000	10	0	
	100		766	166,496		528	83,900

Figure 23.

This represents an actual comparison for a family deciding between an independent home on a mountaintop, twelve miles from town, and a house in town. This site evaluation exercise, like the Tree Game in chapter 5, and like any effort to quantify intuitions or impressions, makes some broad and insupportable generalizations in the interests of practical comparison. The family in the example agrees that a good, solid, five-thousand-dollar fence will be required for privacy and quiet in town. Assuming that the two properties are equal in monetary value, the cost attributed to "Attractive / suitable" represents the extra cost of building a home at the rural site. If they build their house themselves, it better get ten points! The absurdly high cost of driving to and from the rural home includes the expense, at twenty-five cents a mile, of carting the children to school thirty-six weeks each year for twelve years, and of two shopping trips per week the rest of the time, calculated over thirty years; the smaller cost for the urban site assumes that the children only cadge two trips a week to school, and that the family sticks to a twice-a-week shopping regimen. In reality, these trips may cost even more, because rural roads eat cars, and few of us are as good as we should be at abiding by a twice-weekly shopping schedule. Rural costs for water, utilities, and community services represent estimates of infrastructure costs—well and pump, solar modules, storage and power-conditioning equipment, heavy road work, and additional rural insurance costs—over the same thirty years.

This kind of exercise emphasizes several points. In the example above, we estimate that it costs half as much, and will only be a third less satisfying, to live in town, assuming a direct relationship between money and suitability points. And yet, of the five most important elements, the town dwelling comes first only once, and total rural suitability points are twice the urban total. This family, already leaning toward life in the country, decided that the rural life would be much more gratifying and that money, after all, is only money. The site comparison process was worthwhile for them because it gave them a clear and arguably rational justification for their decision.

The Site Rose

Geographic and seasonal factors are critical considerations in choosing a homesite. The sweep of winter storms, the spring and autumn gales, the prevailing breeze that tempers summer's heat, can all be ignored in favor of other preferences, but at considerable cost, for an ill-sited (or ill-suited) dwelling is at the mercy of the earth's forces.

Many modern houses were built on the flatlands because that simplified site preparation for developers in a hurry to manufacture as much

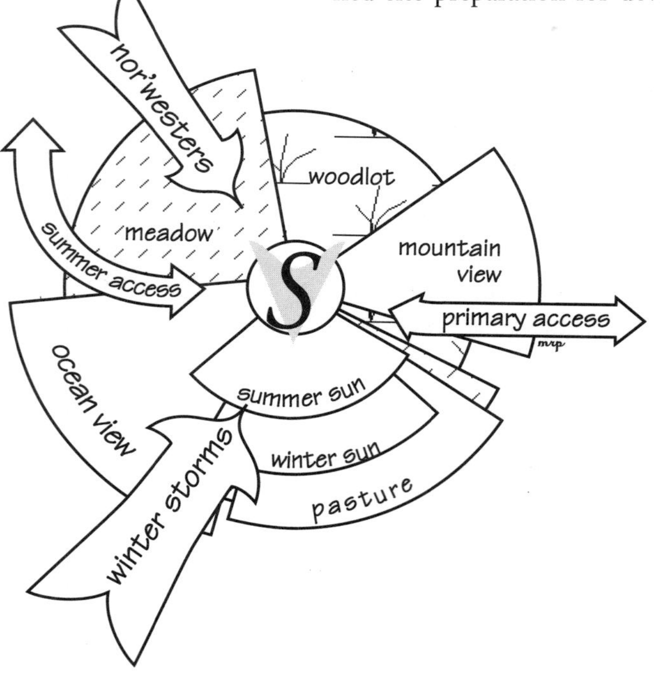

Figure 24. The site rose for my house in Caspar. The best view and the biggest storms come from the southwest.

The author's house in Caspar, California.

housing as possible. Few independent homes are on flat land; their owners explained that they looked to the little hills , not to the cornfields, for their homesites. In the hills, favorable dispositions of sun, slope, wind, and water make a home interesting. A home on a flat site expropriates much-needed agricultural land.

One difference between a house and a home is the harmony (or disharmony) of shelter and site. For centuries, oriental geomancers have studied the *feng shui* of building sites, the concordance between a shelter and the energy flowing in its surroundings. Westerners take similar readings, and call the results their site study.

A site rose is a graphical way of depicting the currents that a site brings to bear on a house. Align a house *against* the currents but instead *with* a street or access road, which may reflect nothing more sensitive than a developer's whim or a cat skinner's convenience, and it will be only by accident or dint of great heart on the part of its residents that this house ever becomes a home. (I hold that love can overcome bad feng shui.) The forces indicated by the site rose are inevitable. Atop my bluff in Caspar, powerful storms come from the southwest and dry, cold northwesterlies cleanse the air two days out of five, all year long. Here as everywhere in the northern temperate zones, the sun rises to the east, arcs high to the south at noon, then angles back to the west. These are the forces to which a house must respond if it is to become a home.

In Caspar, winter's defect is horizontal rain and biting cold, and summer's is fog. Across the northern tier and in Alaska, winter is the problem, with wind, interminable biting cold, and its cohort sleet, ice, and snow; summer's only flaw is briefness. Even if you have never before given a thought to mapping your home's relationship to its ruling forces and features, you could sketch the most basic features of your place's site rose. Some influences exist everywhere, like the sun's light, but others are subtle and may take years and special attention to discern.

Regional Design

There is wisdom in traditional design even deeper than the satisfying coherence it lends to neighborhoods. Distinctively "regional" homes reflect the experience of generations with the weather and the style of life implied by the land in that place. In the twentieth century we have been able to rise above the land by disengaging ourselves from nature, and by massively consuming fossil fuels to impose our own preferred conditions upon our environment. Cheap resources, and our heedless rush to consume them, made it possible to export some strange notions of suitable houses to unlikely locales: sprawling, split-level, ranch-style houses in Connecticut, and tight, compact New England saltboxes in the tropics. But the unfitness of these houses to their new environment will have a direct effect on their longevity and the success of the families who inhabit them. To make an out-of-place house work requires heating, ventilation, and air conditioning (HVAC) equipment reckoned in the tons. Such a structure makes arrogant claims about human dominion, and nature contrives to make such claims costly. It is better to build in accord with the environment of a region. We do better when we bend to nature's powers and adopt the forms that time has proved best.

David Palumbo enjoying the coldest day of the year in front of his shop. The snow-covered modules are not producing much, but the micro-hydro is at its peak.

Improving the land: David Palumbo's story

David and Mary Val Palumbo live with their three children in a comfortable off-the-grid house near Hyde Park, Vermont. Before building, they spent years preparing themselves to build, and years living with their land. Since building, David has consciously polished his stewardship skills, and the land glows in response.

I like doing things myself, and I want to have firsthand knowledge of what it takes to make things work. Technology excites me, not the Research and Development part, but taking proven technology and helping people apply it properly is what I like to do.

I was lucky when I started, because I got some good advice from the pioneers. Peter Talmage showed me by his example how to scale my business so I could make a comfortable living without being greedy, and always take time to be sure my customers are happy.

I got a lot of help on my own house, which is also my biggest and best project, from Milford Cushman, a local architect. I've learned that, whenever I need technical advice, it's best to ask, and the pioneers and experts are very generous. But let me start a little closer to the beginning of the story.

After college—studying business, no science or engineering—Mary Val and I worked for Outward Bound in northern Minnesota, and I learned to love the woods. After a stretch in the wine business in Santa Cruz, I returned to work with my family's real estate business in Massachusetts. It took seven years there to build up my share and get Mary Val through her professional training. When we cashed out, the stake looked small for what we wanted to do in Massachusetts in 1984, so we came up here where land was much cheaper. We knew we were looking for a good-sized piece, with woods. I guess we just assumed we'd be connected to utilities, but when it became clear that the properties that were right for us didn't have power, I set about getting myself educated. We were glad not to add to the need for more nuclear power, and we're individualists who pride ourselves on adventure, so relying on ourselves for our own power fit right in.

We looked at a lot of properties. I often have to hold people back when they move away from the city; few people have the necessary discipline. You don't want to buy land without studying it carefully: you want to make sure you understand about rights of way for utilities, about neighbors, about how the seasons and weather extremes treat your property before you commit. We went slowly, and ended up with a better site than we knew.

We wanted a place with privacy, but close to commerce and Mary Val's work. We wanted to be at the end of our own road, so we could choose our neighbors. We wanted woods, at least fifteen acres, we thought. After we got over the surprise of going off-the-grid, we knew we needed a good balance of energy sources.

We were very lucky finding this land, because we had made an offer on a smaller site, for more money, but this came up and that fell through. This land needed some bushwhacking before we could appreciate it. We knew from checking the soil and drainage that there were two good house sites. We lived in a tent on the best site while we built a small, efficient, quick house, twenty-four feet square, facing exactly south on the second best site, so we could start living on the land. We did a lot right with it, and it's still a very workable place. The PVs are an easy broom's reach for clearing the snow. We learned what we needed to know to start planning this house.

This is a cloudy place, and PV is not enough by itself this far north. Our original concept was to use as much PV as we could and make up the difference with a propane generator. There are two houses and a big shop on our homestead grid, so we manage it like a small utility. We use some big power tools, so we knew we would need a good-sized generator for back-up.

Our favorite time of the year is July and August. There are a few days when it gets so hot, you want to spend time in the pond. That's how we got the idea for hydro, prospecting for pond sites. We found three streams on the land, and one of them runs year-round. Getting the hydro system working was more hit-and-miss than it had to be. I know how to do it now with much less fuss, and I'm looking forward to doing this one over, but it works well. We only use the propane generator when we run the bigger tools, like the planer.

We didn't know how many children we would have when we built this place, so we made it big. All the framing and trim lumber came from the land—we used as much indigenous material as we could. We used a lot of wire, because we wired for 12, 24, and 110 volt.

During the winter, it takes some time to manage all the systems, the electricity, the heater, the children. The kids go to school and childcare, and Mary Val commutes to her work as a geriatric nurse practitioner, specializing in continence care using biofeedback and exercise. It's a new field, and she also teaches at the University of Vermont four days a week. During the week I run the house and do my work helping people put their technology together properly.

The big Essex wood gasifier runs all year long. It burns at about

1800–2000 degrees Fahrenheit; when it's cold out, I fill the firebox two or three times a day, but when it's warm, I fill it once every two or three days, enough to keep us in hot water and the chill off. We thought a lot about the way the systems would interact as we planned and built the house, so there's a wood chute in the garage for getting the wood to the basement and the gasifier. And my workshop, which is above the garage, has a trap door and stairs so we can move equipment up and down easily. The root cellar turned out well, because I can work the variables; many people forget that you have to control humidity as well as temperature, to keep things from rotting. Domestic water is another problem. We have plenty of pressure, but if everyone starts a shower at the same time, the gravity-feed line just can't supply enough, so I've put a pressure tank in for just such times, and now we can hardly tell there's anyone else on the system.

I like the flexibility of 24-volt storage, and most of our appliances and loads like it too—as much, I guess, as 95 percent of the DC loads: lighting, the Sun Frost refrigerator and freezer. The little task lights, draft-inducer fan, hydronic pumps are 12-volt. The washer, dishwasher, dryer, microwave, TV, and VCR are all standard 110-volt AC. I have my batteries laid out as two 12-volt banks. I anticipated that balancing the loads, and keeping both banks at the same state of charge, would be a problem, and I set the 12-volt draft-inducer, the fan that makes the gasifier go, so I can switch it to draw from whichever bank is stronger. I have three sources of power, which is a little complex. Especially in cold weather, generators give off-the-wall power when they start up, both frequency and peak voltage. I learned the hard way to be sure there is no load until the generator's RPMs smooth out.

I studied the way Native Americans, especially the ones who lived around here, treated the land. Flatlanders get emotional about cutting trees, but the regenerative capacity of the northern woods is staggering. You don't need to replant, and you can't keep the woods back. There are two hundred and fifty wild apple trees on our land, planted, I suppose, by the original farmer who cleared it. The forest is crowding them out, but they are important for the wildlife, for bear, deer, and grouse. We've worked with the Department of Agriculture to release those trees from competition to the south, and now we're investigating some edge and patch cutting. Patch cutting (clear-cutting a small area, no more than a half-acre, defined by the forester) can make flatlanders really howl, but it turns out the edge of the regenerating forest, where the poplars start, is a crucial part of the woods habitat, where most of the animals thrive. The young poplars, it turns out, are necessary for grouse reproduction. We're doing what the original inhabitants did, bringing sunlight into the forest.

David Palumbo adjusts the flow to his micro-hydro generator. The valve and penstock are buried below the frost line.

Meeting Our New Neighbors

When we come to a new place, it is wise to seek out its oldest inhabitants, to learn to honor them, in order to appreciate and secure the future of the *genius loci,* the genius of the place. (I mean "appreciate" in both its senses; in addition to enjoyment, wouldn't it be nice if it were our human habit to add value to everything we touched?)

The oldest inhabitants are usually the native plant and animal community, who can testify in enormous depth about how best to live with the land. Where human encroachment has reduced the nonhuman populace to mice, insects, and ornamental foliage, relevant information and experience is lost. In the parts of the world where humans have overpopulated and driven other species away or to extinction, this loss is tragic and often irreparable. Conscious people everywhere are beginning to see (though in many places, too late) that native flora and fauna are inestimably, inconceivably valuable, and that they must be preserved and restored at any cost. When we favor indigenous plantings over exotics we begin to appreciate the importance of an ecology that is attuned to its specific place, and that maintains itself without expensive and labor intensive intervention.

The appearance, habits, and homes of our native, nonhuman coinhabitants yield crucial clues about our chosen place. Citified humans may bring a sort of nature blindness with them, an inability to see a butterfly, to hear a bird. There is a wealth of literature written by people who are recovering from this handicap, and are drunk with the joy of finding that the minuscule and majestic survive despite humankind's supposed dominion. For those whose previous experience of nature has been found in books, movies, television, and the denatured world of modern landscaping (a narrow pseudo-ecology contrived for its visual effect) riotous nature comes as a delight . . . and a shock.

The redwood trees that dominate the forest on my eastern horizon have never colonized the windswept blufftop. The earliest photographs of Caspar, taken just as whites began thrashing through the forests, show a wonderfully thick stand of shore pine right to the rim of the ocean bluff. The shore pine are largely gone, replaced by unsuitable exotics, cypress and eucalyptus, which are not happy. (For more on natives and exotics, see chap. 11.)

Domestic dogs and cats, as out of place as the cypress, have devoured ground-nesting birds and hounded the large game away from settlements. Hooved locusts—cattle, sheep, pigs, and goats—have had their way with the meadows and wetlands, where the tenderest, rarest plants have quietly subsided into extinction without ever being known. The herders, citing the need to protect their valuable livestock and forage, have hunted the wild creatures deeper into the woods.

People everywhere are beginning to long for the confirmation and guidance that only a healthy nature can grant. In Colorado, a jogger is attacked by a mountain lion, and the county animal control officers offer to abate the nuisance; the neighbors, jogger included, resist vigorously, defending the magnificent cat's right of place, saying, this is why we moved out here. "Perhaps," admits the shamefaced jogger, "I should have been paying attention and looking around, not listening to my Walkman, as I ran at dusk."

In California the six years of drought beginning in 1986 forced the understanding that our land is marginal desert, and awakened our interest in xerophytes, plants that tolerate dryness, like the native plants which had been grubbed out and replaced with thirsty exotics decades ago.

What the Woodrat Recommends

Besides the redwood, the most intelligent of my surviving native neighbors is the woodrat, who recommends the following building practices for Caspar:

(1) Choose the south-facing slope above the flood plain and out of the way of rivulets that run downhill in big rains. Do not select a site without surveying it in the wettest part of winter. A spot below the brow of the hill and out of the northwesterlies is preferable; the riparian coppice along the creek is chancy to cross even in the driest years, and most years knee-deep (to a human, to say nothing of a woodrat) from first storm until May.

(2) Build with wood, in layers. The woodrat piles on twigs and sticks extra thick each fall; the heat generated by their composting process will keep the carefully-drained nest, lined with feathers and other found softnesses, cozy on even the coldest night. The entry to the home is through a vestibule similar to that found in New England homes, which keeps the weather outside.

(3) Learn to live with the wet, carrying on life in the rain as in sunshine. Larger species, fox, coyote, puma, and bear, are so rare that we may only speculate how they live deep in the small remaining stands of ancestral forest. If brown bears still survive there, do they hibernate? In winter there are whole weeks at a time when I would if I could.

Hummingbirds recommend migration, and follow their own advice.

Original Dwellings

The white exploiters have managed, in all but the least desirable places, to dispossess the original human inhabitants of the land, but where indig-

Figure 25. From: The Atlas of the North American Indian *by Carl Waldman, illustrated by Molly Braun. Copyright © 1985 by Carl Waldman. Reprinted with permission by Facts On File, Inc. New York.*

enous peoples have lived, we may study their shelters, which in most cases have been patiently refined through the ages to fit the locale. The Yuki and Pomo who occupied Caspar before the whites came maintained two homes and travelled from near Caspar in summer to the inland valleys near present-day Willits and Ukiah for winter. Their comfortable, low, sleeping houses were hemispheres made from abundant, sustainable local materials: rock, mud, willow, and redwood bark. Daylight hours were spent outdoors except in the worst storms. Communal kitchens, bark and leaf shelters without walls, were separated from the sleeping houses, and were focal points for the community's daytime activities.

These masterpieces of adaptation to local materials and climate offer models of careful siting and orientation—of sensitivity to the common forces, summer and winter, and to the basic human needs for family, privacy, and shelter. These original, regional dwellings are most worthy of study. Over and over in my travels homesteaders told me that traces of much earlier occupation were found at the finest homesites: arrow points, stone tools, and the debris of stone-napping. In Caspar, we have found points and a small shell midden. When these ancient conceptions about tribal housing are translated into modern shelters for extended families and unconventional groupings, we often see better adaptations between the people and the land than can be provided by the European-style single-family residence or the conventional arrangement of a community into tracts of houses.

When we listen to the land, and attend to the prehistoric patterns of living in partnership with it, we change the way we live. The land needs us to be ever more attentive and reverent. Living in conscious harmony requires us to actively work with the land. A subdivision dweller is spared

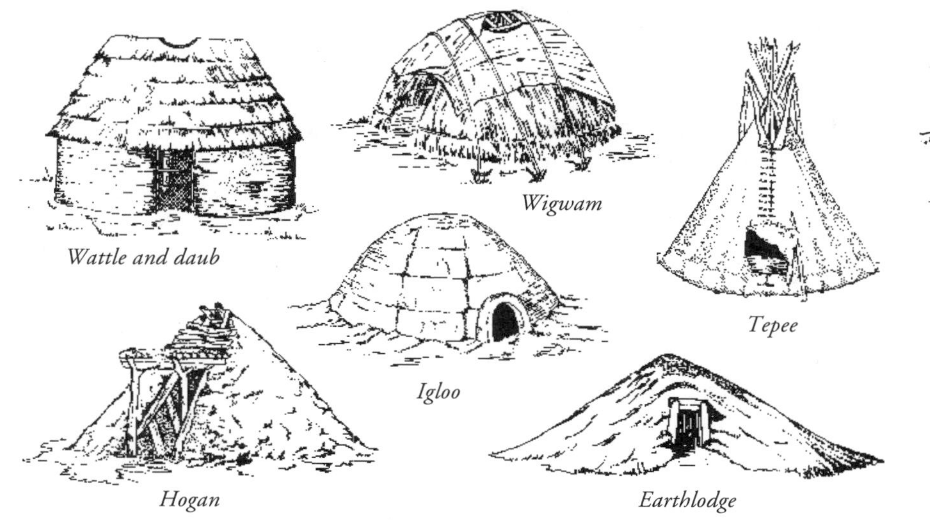

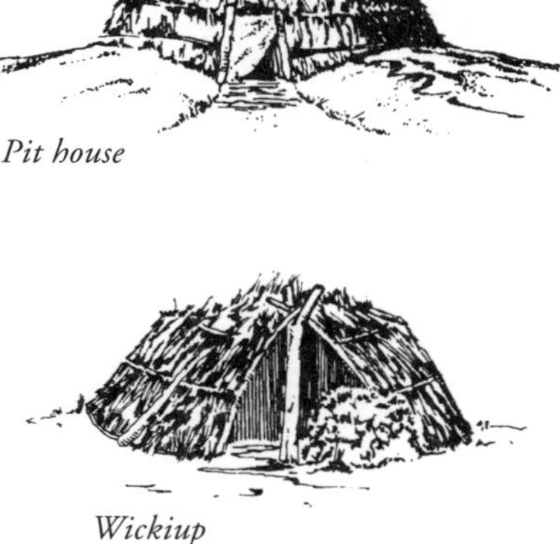

this effort, because he lives in a place where nature has seemingly been tamed (until a hurricane comes along). The land exacts a tribute and dedication from its partners that stretches the conventional isolated nuclear family; just as it takes a whole community to raise a chile, it takes a small tribe to keep a homestead. Cooperative living schemes, shared ownership, road associations, and other adaptations to the needs of the land are leading pioneers to reinvent some of the social solutions perfected over many centuries by previous human inhabitants. We may benefit enormously by deepening our study of their ways.

After seeking the advice of the original inhabitants, we should turn to living masters who have been with the land all their lives. The wonderful insights of these people can bring us closer to the realities of our home place. They are wise about the weather: a cloudless azure January sky in Vermont is a breeder sky which tells us that the storm is coming soon, but clouds coursing in from the southwest promise a storm in Caspar. Knowing these signs allows us precious hours to prepare. The old ones are also wise about shelter. Many old-timers still remember, and will tell us, how houses worked before indoor plumbing and energy dependence changed our perspective; hearing their stories, we may re-evaluate our assumptions about what makes a good and suitable home. In Caspar, where the storms come off the ocean, old houses have few windows because conventional wooden windows leak water and cold air; magnificent as it is, the ocean is identified with the source of domestic discomfort. We might wish to preserve the ocean view, but we know from the experience of our predecessors that we must take great care to make those ocean-facing windows strong and weatherproof.

The Laws of Physics: Laws we can live with

If, after consulting the original inhabitants, animals and humans, we need further guidance, the laws of physics can be called upon to help. All our various instructors may suggest the same solutions. For example, where cold is a major factor, thermodynamics dictates that the best shape contains the most volume while exposing the least surface: a sphere. Spheres are hard to build using wood and glass, and humans require floor-space with headroom, which is a little different from simple volume, so a rough rectilinear approximation of the sphere, the cube, fits us better. In cold climates, successful wooden houses tend to be cubes. Another example: Animals that do not migrate either take to houses themselves (hibernating in caves or tree trunks) or are superbly hirsute. My cat's winter coat makes him look twice his actual mass, and arctic foxes have even better insulation. Cold-weather creatures are usually smaller and more compact than their temperate relations. So it should be with cold-weather houses, which need abundant insulation and adequate but not excessive space, because the luxury of wasted space is absurdly costly. Taking the physical effects of prevailing winds, seasonal waters, and the warmth of the sun into account provides a theoretical model of a house's mechanics. The classic New England saltbox is wider east-to-west, which maximizes solar exposure; it is lower on the north side, so it hunkers down before the prevailing northerlies; at its heart, a massive fireplace keeps warmth streaming outward from the protected center. If southern windows allow sun to warm the fireplace's mass, the heat will reradiate in the evening and reduce the amount of wood required to counter evening's cold: a simple application of the greenhouse effect.

Heat most often comes to us as light in the form of infrared radiation, which travels in straight lines from its source; once it strikes something—a surface or an air molecule—it warms the surrounding air, which convects or rises, forcing cold air to fall. In a saltbox, the daytime living spaces are downstairs, where active occupancy makes cold tolerable, and bedrooms are above, warmed by the rising warm air; doors at the foot of stairways keep the heat downstairs during the day.

Other physical phenomena inform regional shelters. Where summer sun causes too much heat, overhangs take advantage of the sun's seasonal path, letting in more sun in winter and less in summer. Deciduous plantings help by blocking incoming radiation in summer, when they are in full leaf, but admitting more light in winter. In the sticky tropics, where seasonal variation is minimal, the earth's surface is often moist from daily rain and inhabited by insects, because breezes do not penetrate the thick foliage; above the canopy, breezes at the treetop level promise the only relief from the humid heat, and it is sensible to build living spaces on stilts, aligned with and open to the slightest breezes.

Low-profile: Carl Bates's story

Carl Bates lives with his wife Tracy and their two children in an off-the-grid, slate-and-wood, multilevel house. During the many years he lived with the land, gradually digging his house out of the slatey hillside of Prickly Mountain above Warren, Vermont, Carl thought about how a house could nestle best among the woods and rocks.

I'm amazed that people put up with ugliness and ineffective stuff; I guess they're programmed. I like to think, if they see something that works better, they'll want to use it. I don't have a clear vision myself, but when contemplating where I wanted to live, after graduating from college, I could see it wasn't the regular place, but something like this.

There is no such thing as off-the-grid: everything you see here except the rock got provided in some way by an economic network for which we should be grateful, because without its economies of scale, we would have very little.

Carl Bates and kin.

I'm trying to strike some decent balance between technical achievement and independence. To live like this, you've got to be committed to being different. I don't champion anything, but I think one answer to our messy mess is, what works is what's good. There are people living around here in western ranch-style houses that don't fit in, but if they're happy people, that works for them. They've found that critical ability, to be able to appease themselves and feel like real humans . . . and I think we have, here, too. I think the main enemy is a sense of alienation from self, a feeling that life is not what we want it to be.

I grew up comfortably in New Jersey, but I didn't see that money helps. It's more important that our lives be expressions of what we want and don't want. I have to admit, I'm not a good finisher, and I guess I will always be working on this house.

I started this house in 1978 by excavating into the hill slope and using the slate to build the lower walls. Where I hit the ledge, I just left the big, hard pieces, like under the sink over there, so the hillside designed the house. I just followed my nose. At first I lived in a tent, and wintered in Costa Rica, then moved into a trailer when we married and had a daughter.

I've got two girl children now, and they're both readers. The eleven-year-old likes the social life at her school in town, and doesn't like the long walk up the hill. She spends a lot of time in her own imagination.

Snow insulates Carl Bates's house, and the undisturbed woods protect it from wintry gales. Only windows and chimneys rise above the snowy hillside.

We started out growing more of our food, but now we grow more flowers. All our sugar comes from the maples. The community owns the meadow beside us, and I keep two young oxen there; I liked the way they worked in Costa Rica, and I think they'll work as well here.

I wanted a low-profile house, that wouldn't be seen to clutter this wooded hill. I gave up the view, because views work both ways: if I have a view, I also spoil someone else's view. It makes me happy, to walk in the woods ten minutes to get to a knoll with a view.

The house is bermed into the hillside, and I built a strong roof. When I'm convinced I've got forty years of longevity to my roof, I'll add sod. The cold, wet uphill side of the house has a basement behind it. At the top of the house there's a sun room that works well even in winter, and we close off parts of the house depending on the weather. At the bottom by the entry there's a furnace under the floor, which radiates heat. A coil in the firebox warms water for the thermosiphon system that heats other rooms, and the flue runs up under the ramp, beside the hydro turbine, so I capture all the warmth I can. I guess there aren't many houses with an indoor pelton wheel, but I've got it sound-proofed so you can hardly hear it. The stream dries to a trickle for a few weeks every summer, and we do without electricity or run the generator when we need power tools. In the winter, we have eighty to a hundred gallons a minute, so there's almost too much power.

Besides being almost half a mile from power lines, I like the independence of renewable energy sources.

CHAPTER 7

PLANNING A NEW INDEPENDENT HOME

REGIONAL FACTORS DICTATE A GENERAL THEME, BUT THE WAYS IN WHICH WE tune our home to its site and surroundings result in particular, incomparable harmonies. A local hill, a watercourse, or a stand of trees will shift the microclimate slightly, and every site is unique. Moreover, microclimate varies surprisingly in a short distance. Harsh winter winds and unforgiving summer sun may make one spot a hellish building site, yet a more clement one may exist just a hundred feet away. We can discover such things by building wrong, or we can ask knowledgeable locals. The best way is to live long enough with the specific place on the land to learn its music thoroughly.

The land offers valuable features—hills, slopes, streams, sensitive habitat—and provides building materials—rock, timber, clay—that suggest the best site and building method. These contributions are less obvious and insistent than weather, and can be welcomed into the design of a house with pleasing results, or ignored. However, by taking the time to learn the ways of our site, from the imperatives of its landscape and weather to its gentler singularities, and by finding ways to integrate this locale's gifts into our new house as we build, we help the home take its place in the larger orchestra, so that all the instruments may play together.

The house will also create internal microclimates of its own. By anticipating this, and tuning its shapes and sequences to the site and to the rhythms of our own lives, we improve the way the house works for us, and with us. The kitchen, for example, is a net heat-source and is usually occupied by active people, and so will be comfortable at a lower ambient temperature than, for example, a dining room or study, where activities are more sedentary. There is reason to believe that sleeping areas are healthiest if they are relatively cool, but bathing areas, located just next door, are unwelcoming if cold. The passive forces that move heat around within a dwelling—radiation and convection—are predictable, and good lanes for internal ventilation and low-energy devices such as fans can nudge the currents along. The living spaces therefore may be intelligently arranged within the building, until a well-designed living space plays in close counterpoint with the site rose.

The home and its occupants are a sundial. In the morning the sun rises in the east, and most residents arise from their beds and perform their morning ablutions. As the sun moves up across the sky and down again to set in the west, the residents move from bedroom to kitchen, through office and workroom to dining and family rooms, finally returning to the bathroom and bedroom. A house works best when these movements of sun and people are coordinated. Morning light in bedrooms and bathrooms predisposes us for a wakeful day; evening light in the living room helps return us to our center where we may give up external cares and prepare for restorative sleep. The farmhouse kitchen acknowledges that the home-maker's activities (and many of our most enjoyable times) revolve around food, and the kitchen is at the heart of all. Where the kitchen is the sunniest, cheeriest room, a house is well on its way to being a home.

The idea of a house has been steadily inflated and become grandiose under the persistent urgings of Hollywood, architects, and a consumer economy. Dramatic contrasts and too many specialized rooms may obscure the real functions of a home. Do we believe that a house with a neatly delineated kitchen, breakfast nook, and dining room will make a family happier than one with a homey, sprawling kitchen where all those activities take place? When planning an independent homestead, we want our creations to work for us the way we are, not the way our culture tells us we are. In planning a new house, we have an opportunity and responsibility to build as an expression of ourselves.

Many independent homes show an affectionate and humorous recognition of the uniqueness of their sites and their builders. For independent home builders, uniqueness may be an absolute requirement, even if eccentricity presents an insoluble conundrum to the real estate establishment, which prefers comparability and "curb appeal." How can a property be evaluated if it cannot be compared to other, similar habitations in the immediate neighborhood? Independent homes are customized—comprehensively, elaborately, and lovingly. A conventional house's minor uniquenesses, kitchen cupboard doors or the tiling in a shower, are described by a real estate agent as "charm." By that estimation, alternative homes are all charm, and completely incomprehensible in the real estate context. Early in the planning process we must accept this gamble: with luck and infinite care we may build a perfect home for ourselves, but its perfection may never be appreciated by anyone else.

The biggest challenge in building a home that suits us is to step far enough back from conventions, habits, and expectations that a vision of our real needs, and the site's special qualities, appears clearly to us. Every house has good and bad aspects; the idea is to build a new house that borrows good ideas from many homes, balances and integrates them well, and keeps the undesirable features to a minimum. Under time pressure, and pushed by financial realities, we often settle for old solutions and wrong assumptions, thus compromising our vision. If we intend to live for a very long time in the home we are making, we should not rush, but patiently wait, work, question, and plan until we find a way to accommodate reality hospitably within the vision.

Here is an example: Glass is transparent to sun whether perpendicular to the sun's rays or to the floor. Popular solar architecture proposes dramatically angled glass, slanted perpendicularly to the sun, aimed like a solar collector. And yet angled glass walls reduce headroom and volume; their glass is subjected to unusual stresses, dust, and streaking; framing and

Figure 26. Sloping glass costs headroom and complicates framing.

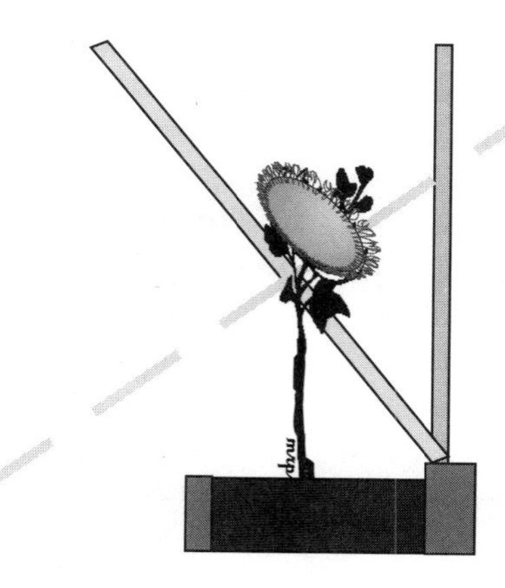

weatherproofing slanted glass walls is more difficult, more expensive, and requires more maintenance. It looks modernistic, daring, space age; experience unequivocally proves that it does not work as well as vertical glass. Arnie and Maria Valdez fell into this trap when building their house; now they plan to rip out their mistake and rebuild their greenhouse walls vertically, creating more space in the greenhouse and on the floor above.

Comfort apparently seems unnatural to home designers, but concern for comfort comes quite naturally to cats, who spend all day, every day following the trail of sun through the rooms of the house. Possibly the element missing for designers is time: they never spend enough time in one place. Too often, houses by architects are complicated expressions of an idea about a site, which work very well one or two days of the year and are wrong every other day. When the design for a home evolves out of sensitivity to the qualities of a specific site and the needs of an actual family, that design is likely to work well most of the time.

If cats designed homes, this is the way they would go about it. In my home, windows, walls, and archways have been built to welcome sun and to shun storm on behalf of the space within—the way a plant grows, with tougher exterior and protected heartwood within. My cat approves of the result, while architects scratch their heads. Since storms come at my house off the ocean, windows in a roof (as on the hull of a boat) are a challenge. Even as the storm breaks against the windows on the prow of my house, the ocean view is striking, and I could never give it up. But I have not yet found a way to keep the weather out; instead, the most exposed windows are over greenhouse and bathtub where a little indoor rain can be managed.

The bathtub occupies the loveliest spot in my house for a sunny winter afternoon, when you will always find my cat napping on the ledge beside the tub. Living in a conventional building near this site for five years before starting to build, I watched the cat. She always knew where to find the best spots for any time of day in any weather. Our bitter, wind-blown storms, so intense that southwestern walls must be as waterproof as a roof or a hull, render even the coziest places (like under the bed in the northeast bedroom) barely tenable. During the storm, what better place to have a leak than over the bathtub? When the storm has passed, what better way to celebrate surviving the storms than in the bathtub, with my face in the reborn sun, sharing the solarium with the cat and the first daffodils?

Living grounded: Hamid and Saqina Bush's story

Saqina and Hamid Bush live in an earthship built by Greg Martin and Jenny Bird next door to their own pit house on the mesa northwest of Taos, New Mexico. The idea for the earthship came from Michael

Reynolds's work (see chap. 8 and 14), but they built this one to suit their own needs. During the day Saqina and Hamid teach at Chamisa Mesa School, a private alternative secondary school, while the earthship takes care of itself.

Saqina: Now that I have lived in an earthship, it would be hard to give it up. The amount of sky is the best surprise. It would be limiting to live in a room with little tiny windows.

Of course, now that I've lived in one that somebody else built, I want to build my own. This is Jenny Bird's fireplace, and she and Greg laid out the space very comfortably. But it makes me want to enjoy space I've made.

Hamid: This earthship is built in a hole about four feet deep at the front [see figure 31]. When you stand at the sink, your eye level is just above the level of the plain, and so you look out continuously across the mesa to the mountains.

Saqina: The sloping glass lets you stand with the mesa in front of you, and sky right above. It makes a very special space.

Hamid: They dug the hole several years ago, and the weather since accounts for some of the soft roundedness to the earth walls. The finger walls and bathroom wall work as thermal masses, but the floor has not reached thermal equilibrium yet, and has never been warm to the feet. It's dry now, but was real damp when we moved in . . . two days after the floor was finished. I think a closed-in summer will take care of this problem.

We helped them finish the inside walls . . .

Saqina: I got to build one of the adobe brick walls. It was very simple, one of the easiest things I've ever done. And I painted the white parts inside the house. Jenny and her sister Tracy painted the colors. At first, when I saw the orange and blue in the bathroom, I said, whew, color! But now I like it.

When they built the first fire in the fireplace, they almost burned their roof down. I guess they missed a spot in the chimney, and the smoke was going into the roof. That was a troubleshoot!

We moved in when the house was still raw, things like no covers on the electric boxes . . .

Hamid: Like moving into a new house. It makes us really conscious of the house and what it's doing. And we are constantly conscious of the earthship doing something, like it's alive.

Before living here we lived in El Salto, in a pumice house (a mixture of very light volcanic rock, a great insulator, and cement, then poured like concrete, sixteen inches thick), with a slab floor. It was a good

Hamid and Saqina brace against a winter wind as they leave their earthship on the way to school.

Morning light sparkles through the bottle walls above the shower in the Bush's earthship.

passive solar design and very easy to heat, but not as good as this earthship.

We've wintered here, but haven't summered yet, and it will be interesting to see how that works. This house was just closed in, in November, and so it hasn't gotten its thermal masses charged up yet. I guess it will take a full summer to do that.

In the early morning it normally gets down to 62 degrees unless you have a fire. As soon as the sun hits the windows, it starts to heat up. When it hits 80, you open the end windows and it stays between 70 and 80 degrees until the sun goes down and it starts to cool off.

Hamid: In the past I've always liked warm houses, but now my body is quite comfortable with 65. My body has adjusted its comfort zone by about five degrees on each end since we moved into this house.

One thing that's very different from a conventional house: there's a different relationship between the walls, air, and floor. When the sun goes down, you can feel the walls radiate heat, just like a fireplace. But at a certain point in the evening, probably determined by the amount of sunlight we got during the day, they start to radiate cold. The thermometer may say 72 degrees, but here in the earthship, surrounded by stone, plaster, cement, and tile, and all colder, it feels like heat is drawn directly from your body, not just the air, and so you feel colder. You can feel a suck happening, but the sensation of the back wall changes a lot. It's demanding heat. Or radiating cool. What a bizarre phenomenon, radiant cold! After a summer enclosed, when the walls will be charged up, I think the radiant heat will overcome the radiant cold.

The earth buffer effect is very interesting. The earth's temperature remains at 52 degrees, I guess, and so you have to be careful of your glass to mass ratio. In winter, the ground wants to cool down, and the outside temperature may be below 30 degrees. It's only been above 50 outside once since mid-November. So for five months the earth has the advantage over the air here; the sun through the glass and our wood fire are the only way to make the gain needed to be comfortable.

There are good feelings that come from living partly underground, as if it protects us from an energy that penetrates above ground houses. I rest well in this house.

The earthship's interior space becomes very personal, determined by the way a few things are placed—the end walls, for instance—and by how the front communicates along the window. Any wall near the front window makes a big shady area.

One of my favorite things about this house is the greenhouse all along the front, a counterpoint to the rest of the house. I imagine building a nice long earthship with another row of planters along the front, a wandering place inside your house. In the wintertime it's too cold to go outside and fart around. Maybe you can, but you don't want to be out when it's so cold. So I come around to the idea of a larger greenhouse inside so the front of the house is a tropical jungle.

Having the big open studio changed my idea of the proper size for rooms. When we build our own earthship, we'll make a space like the studio, because it's so glorious to move around in. With the windows, and the way the front of the house is just above eye-level, there's a sense of a day at the beach. The light is that same kind of ocean-open light.

Saqina: There is so much light coming through the sloping windows that everything gets bleached. You can't leave your bathrobe on a post in the front of the house for a week, or it will be ruined. And the plants get so much sun, I have to water them a lot.

Shades, maybe even automatic ones, would be a good idea, and ventilation windows in the end walls that open automatically when it gets to 80 degrees.

Hamid: When we think about building our own earthship, we both like the idea of the planters along the front, and one or the other end of the house becoming a complete indoor patio, with gravel floor, an indoor/outdoor space . . .

Saqina: . . . with straight walls and doors I could slide, so there would be one place in the house I could look straight out. The hot tub and shower would go in that area.

Hamid: The middle of the house would be the living part, work space all on the inside, then a bedroom at the end opposite the outdoor space. The bedroom might be the only separated-off space, but all the rest would be connected.

Except for water-based paint and cement dust, the house has no premade materials, nor glues or solvent-bearing materials. It's basically mud and wood and rock. The house is vapory neutral, which an all-wood house isn't: I always have a sense of glued and resin-impregnated material. In the earthship, you're in the earth.

Almost everything except the glass comes from close by. You can get a vega permit from the national forest people, which lets you go into the hills and cut standing dead trees for the main roof beams. Between the vegas you can put latillas, like straight branches; this is the way an adobe house used to be built.

The tires are interesting: cheap, labor-intensive, crude bricks. It's like building a cave. I'd definitely get somebody else to pound the tires. It takes a lot of brave young gentlemen and ladies who have the physical energy to do that.

I imagine it's a real thrill to build a different kind of house. This house is really a good idea. But it is a cruel thing that you can save up $20,000 and buy five acres of mesa land or higher-up land, and contract to have a $100,000 adobe box or pseudo-adobe stucco finished off, and walk into the bank and the bank will give you $80,000 and a regular mortgage. On other hand, if you've got only $10,000 and want to borrow $70,000 to buy two acres, and you take in the plans for an earthship, and everybody agrees they're fine, and you've got your crew to build, the bank would say no, you can't do that. You can't get a construction loan for an earthen house. Not because of building regulation, just a flat banking decision somebody made.

We wonder what it would take to bring in an earthship: maybe forty dollars a square foot, which is relatively cheap in a weather-intense place like Taos . . . if you were willing to work it yourself just as hard as the crew, maybe fifty dollars if you were just honchoing the job. So you're talking about a two bedroom home on five acres for under $100,000. I think these homes, if well made, will always be saleable.

Here, our storms come in from the south and southwest . . .

Saqina: The lightning storms in the summer are awesome. In October the sunsets are magical. One was so beautiful it made me almost cry. The clouds were purple, with a metallic golden quality to the west, and the snow on the mountain had a peachy purple quality.

Figure 27.

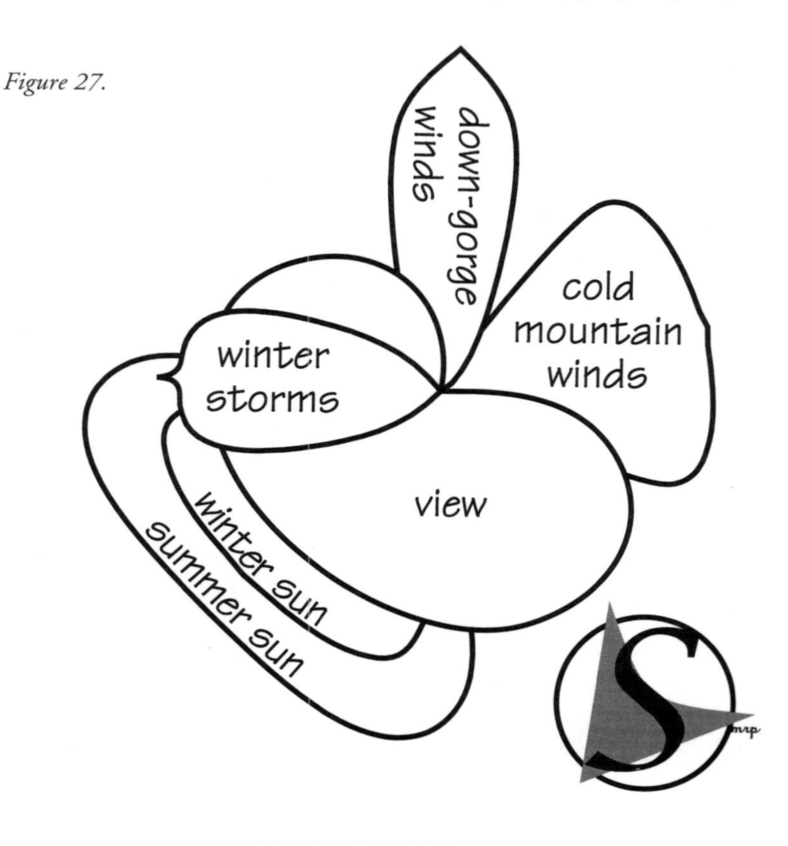

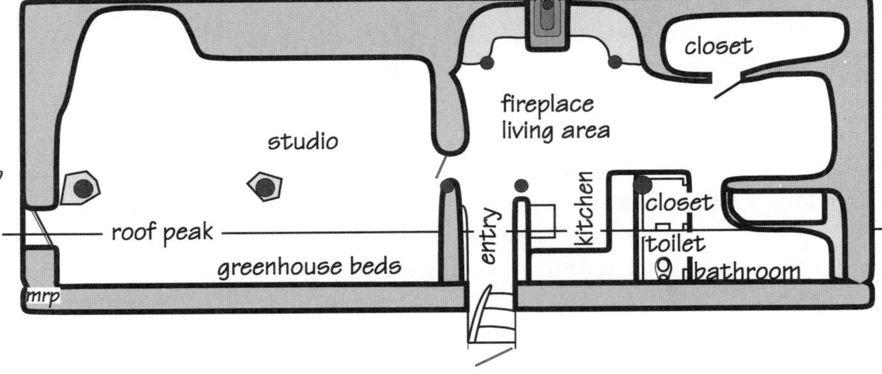

Figure 28. An earthship's back wall is sculpted out of the earth itself. Other walls are built from earth-filled tires, bottles, and cans covered with adobe.

Hamid: When it snows, the windows get covered, and it gets totally white, eerie, translucent white. Cold is really clean, because everything goes away. White snowy ground, and black winter shrubbery is visually clean, and there's nothing but air in the air, not even water.

Saqina: It's insulating. There's no privacy except on the snow days, which feels neat, because I'm not in the fishbowl.

Hamid: Cocoony is a good word.

Saqina: A strong wind comes down the gorge in March and on through the spring during the day, not so much at night. There's no consistent or predictable wind from July to March, two-thirds of the year. But if you wanted to move storage water in the spring, you could do it with a windmill.

June and July and are the rainy time, and July is our big rain month. It rains for a little while, then it's sunny, and so PV could also be very effective at that time for moving water. This is a high desert, and to grow much of a garden, you'd have to store a lot of water.

What Can We Learn from Existing Homes?

The modern American home is built like a space capsule, tightly sealed to conserve heat and utterly estranged from the outside, its occupants viewing nature through large windows as if on big screen TV. Within such a house we may, if we like, ignore trivial issues like the sun's rise and fall, its seasonal declination, and the march of the seasons; these are dealt with by the big HVAC (Heating, Ventilation, and Air Conditioning) unit. A conventional home maintains the pretense of human dominion over nature—at great expense.

To those who have chosen to live in alternative homes, this seems an arrogant, costly, unsatisfying, and unhealthy circumstance. We seek ways to connect ourselves and our living spaces to the pageant of weather and seasons, to involve ourselves in flux rather than negate it with expensive and wasteful technology.

The conventional house's rooms are most often arranged for speedy construction and maximum profit for the developer. The kitchen and bathroom are likely to share a wall, not because there is inherent sense to the arrangement, based on the dynamics of the family that will live here, but because it is cheaper to build if all plumbing components are consolidated in a single wall. Reckoned over the life of the house, how much more would it cost to optimize a house's design for the sake of the long time it will be lived in rather than the short time it takes to build it? Modern homebuilding, a form of commodity production driven by the developer's profit and convenience, considers irrelevant the idea that the product must finally serve the customer. It hardly occurs to the inhabitants of a subdivision dwelling that criteria they use to evaluate the merits of a microwave or automobile, that is, suitability, ease of use, and effective function, might be applied to their house. As a direct consequence, we have become a restless culture. When we at last recognize the unsuitability of a house, we simply move: the average American family moves once every seven years.

Independent home builders often revive old building practices and approaches, and work within time-frames forgotten by the modern construction industry—and we independents are often frustrated by the "housing" industry and its building codes. Conventional residential construction (as distinct from fine homebuilding) was invented during the building boom that followed the second world war, when it became a national credo that every family could aspire to own its own home. In the interests of speed and profit, the new industry took its lessons from mass-production of consumer goods instead of from the artisan homebuilding tradition. Modern building codes supposedly responded by guaranteeing the economic soundness, as well as the healthiness and safety, of any house; the results of Hurricane Andrew make us wonder how well an industry can regulate itself. The charmless prehung door and aluminum-framed window are symptomatic of an industry compelled to create space so fast that there is no time for aesthetics or durability. A modern house can rise from a bare site and be occupied in barely thirty days. At the beginning of the twentieth century, a house took at least a summer to construct, and even then only the essential parts of the house were finished, with the remainder waiting for other summers to be added. Houses usually grew as families grew.

The built-in cabinetry in a turn-of-the-century house was carefully matched to that structure and that resident. Modern, prefabricated cabinetry is modular, often imported from another continent and made with materials imported from still farther afield: Danish cabinets made with wood torn from the Southeast Asian rainforest.

Right Ingredients

There is something atavistically satisfying about building with materials taken from the building site. The timbers in my house, milled by my hand from storm-felled trees from the forest just a short walk to the east, make me happy in a way that is hard to explain. The earthships and adobe house of the southwest and the stone houses of the northeast have the same homey resonance. Economics aside, building materials from the site build the best homes.

Building a house is a once-in-a-lifetime invitation to meditate on how we conduct our lives. If we speed through our busy days, we seldom take time to notice the numerous adjustments our homes impose on us. Most existing houses are old or commodity housing, where original intentions are obscured by subsequent alterations—or were conceived in the first place only to be quickly erected. Most of us have lived for generations within unsympathetic walls, and have developed a thick insensitivity to the inadequacies and indignities to which housing subjects us. We do not expect a house to care for us well. Even as investments, houses are poorly evaluated; the real estate machinery makes every effort to keep buyers from talking to sellers about the true qualities of a house changing hands. And we have no vocabulary or procedure for investigating and expressing how well we are suited and served by the place we call home. Yet we commit so much to our houses—our money, and so much more that we value: our time, our love, and our family—that we may tend to turn a blind eye to a building's faults and unwelcome qualities.

Independent home builders working in recent decades have had a chance to put prejudice and defensiveness aside, to formulate new ideas and try them out. As you would expect, you'll find in looking at the homes of pioneers that some of their ideas work well, and some are wrongheaded. We, too, have invested so much in our homes that we are reluctant to admit failure, but at the same time, almost every independent home builder I met offered some advice about how to do it better next time. And because these houses are ours, built with our own hands, we may have no compunction about ripping out wrongness and trying again.

In chapter 5, while considering the cost of demolishing a structure, I noted that building materials often outlast structures. Glass, copper, timbers, and fixtures are not exactly indigenous materials, but are righteous ingredients which can find useful new lives in independent homes. As I built my house, I mined wood as I disassembled the old Caspar Cookhouse, a barnlike structure rotting with disuse in a nearby cow pasture; the posts holding up its porch roof, hand-carved from straight-grained, first-growth redwood are as sound today as they were when milled a century ago. Now they support my stairway and roof timbers. Architects, contractors, and building inspectors all perpetuate a wasteful cultural prejudice against using old materials in new houses. It takes longer to use salvaged materials: nails must be pulled, unusual dimensions adjusted to, plans changed to accomodate found objects. On the Hawaiian island of Kauai I saw acres of trashed building materials, the aftermath of Hurricane Iniki, but despite scarcity of new materials and scalping on the part of suppliers, residents were forbidden to search the dumps for materials they badly needed. Here is another case where near-term cost can no longer be allowed to justify long-term wastefulness.

In planning, we hope to learn lessons from others who have gone before, so we may build our own home "right" the first time. In the pages that follow you'll find some pointers and details passed on to me by independent home builders, discoveries validated by my own observations and experience.

Comings and Goings

Our attitude toward a home is prefigured as we approach the house and enter its space. The threshold that makes a transition from ground to floor may seem difficult to imagine when planning, but because it is so decisive in establishing our experience of a house, the entryway deserves special attention. In my own home, I mistakenly left the entry functionally unfinished for years, until a beloved guest slipped on the milk crate which served as the step up from the mudroom and twisted her ankle. Before that, I had hardly noticed the awkwardness, nor had family members complained. When we move through doors, we are generally moving fast, and are often burdened and preoccupied, and habit is accompanied by insensitivity to any inconveniences, no matter how irritating, until we suffer a mishap or a visitor calls our attention to an overlooked problem. We were all amazed by how much more pleasant it was to enter through finished space, stepping on real steps.

The area around the entryway is a natural collecting point for packages, tools, and other gear which are in transit, are used only outdoors, or have no proper place of their own in the home. Furthermore, when we enter from outside in a storm or during mud season, we bring with us hitch-hikers we would rather not bring all the way in. Something must be done to control the clutter, and a mud room or vestibule well-endowed with storage, counters, and a place to sit while unbooting seems the obvious answer.

More and more homesteaders, especially those living in muddy, snowy, or dusty places, shed their shoes at the door. In Hawaii, where this is standard practice, a small set of shoe-sized cubbies on the lanai keeps things organized.

The entry's placement should be determined by the site and by the house's floor plan; the entry should be from the calmest side during the worst weather, to protect the home's microclimate from the brasher environment outside. In Caspar the protected east side, which is seldom assaulted by storms or prevailing gusts, is the correct side for an entry. If we insist on placing the entry in the teeth of the gale, we must use an air-lock style vestibule, or risk rearranging papers and other light objects on surfaces near the entrance whenever the door is opened. Even the hinging on the door is worthy of our attention: Entry doors traditionally open inward in greeting, but to right or left? Weather, traffic patterns, and even the predominant right- or left-handedness of the inhabitants should enter into your calculation. I decided to flaunt tradition and open my front door outward, thereby preserving precious indoor space; it places new guests at a slight disadvantage, but friends learn to come right in.

Our culture has brought its infatuation with cars all the way home by putting the car before the house. Houses are quite commonly oriented to street and driveway rather than to sun and land. By combining garage and living space, we further increase the distance between ourselves and nature. We have made it possible to live as if we were on a hostile planet, where survival is possible only if we never leave the protection of house, car, office building, and shopping mall. The homiest houses I visited severed this unholy connection by conspicuously excluding vehicles and all their drips and smells from the dwelling's environs. For example, at Frank Dolan's home, which doubles as the parrot hospital, cars are parked well away, because they upset the birds. Years ago, I elected to build my home one hundred meters from the street, because I wanted to preserve the greenery near the house, and keep away from the noise and smell of autos. I expected my guests to be inconvenienced, and put in a lighted walkway to guide them home, but I have been gratified to hear them comment that the short walk gives them a chance to compose themselves and enjoy the garden and view as they approach. Another unexpected and beneficial effect is that we plan our shopping with a bit more care, and travel with fewer bags, because we consider the distance from house to car as we pack or purchase. We, too, appreciate the insulation between our home and the bustle a hundred meters away. Awkward, necessary items come in by wheelbarrow, and the house may be approached across the meadow by car or truck for major deliveries. Having disassociated house and vehicle, I can think of no reason to go back, and my family enthusiastically concurs. The vehicle and the home seem to be uneasy partners, and a little separation seems to work well for both.

Framing Pictures with Windows

When I built my house I was so preoccupied with passive solar heating that I missed an important source of pleasure: nature's clarion wake-up call, the morning sun. In Caspar the daybreak seldom contributes much heat, but introduces great optimism and cheer. People often employ heavy drapes in their bedrooms to keep out morning sunlight, preferring to awaken themselves with alarming clocks. As an exercise in subtlety, I invite you to set off your alarm clock; even the way we say it, with the same words we use for detonating a bomb, warns us what to expect. Can we possibly arise well-disposed toward the day after such a greeting? An alarm clock is a personal version of the demon mill whistle that roused our ancestors in their company houses in their company towns. I find it so much better to have a morning sunray, or even the stormy day's pallid light, summon me from my beloved bed.

In too many houses, window placement is an unconscious act. There are two important issues to consider: what happens inside the house beside the window, and what is outside the window. When planning a house, careful comparison of floor plan and site rose informs us where light will be required inside and where the grand views will be. Often the only window we get right is over the kitchen sink. The standard architectural solution, the view window, is costly to install and more costly to maintain, because even the highest-tech glass is twice as wasteful with thermal energy as a standard wall, and a single pane window is fifteen to twenty times more wasteful. What folly, to put in a big, expensive window then cover it with expensive thick drapes to keep the heat in; we give up the wall space to the view, then the view to the drapes. Conscious placement, where we visualize each window's view corridor—the whole sweep from the horizon and interesting distant objects, through foreground landscaping and trees, through the window and through furnishings to the place where the view will be enjoyed and the outside light employed—is a difficult art, but worth the effort. When a window is placed precisely, it may be quite small and still do its job very effectively. We willingly spend thousands of dollars on wall adornments, but what adornments could be more uniquely ours than those living prospects framed by the windows in our living space?

These southside windows yield powerful quantities of sunlight, and offer an excellent view of the PV array on its tracker.

Conventional houses seldom place skylights for nighttime views, but independent homes often have moonwindows and star viewports. My daughter's bed, at the prow of my third floor, has two skylights and an ocean window. On a clear, moonlit night it is cool there because of all the glass, and the moonlight is astonishingly bright.

Before You Build: Wes and Linda's Advice

Linda and Wes Edwards moved to their country with their daughters two decades ago. They now live in one of the most gracious and well-conceived independent homes I visited. I asked Wes and Linda to summarize their strategies. (You will find Linda and Wes's story in chap. 12.)

Wes: Okay, it's time to take some notes now! Here's my design strategy for a self-sufficient home. Start with a small house. That minimizes the impact on the site, and takes less materials . . .

Linda: When I clean this "small house," it seems big to me.

Wes: And when the grand-kids come, it would be nice if it were bigger.

Buy a plan for your house, and get an architect's help to make the changes you need. This can save you time and costly mistakes, maybe irreversible mistakes. I definitely made mistakes, so I know. I'm an electrician, not an architect, so there were things that I didn't understand that an architect sees every day. It can be done right the first time. We spent two years seriously planning this house, gathering information, studying plan books, visiting other people's houses, watching the site through the seasons. From groundbreaking to occupancy only took a year.

As you build and change things, keep the house plans current, or keep sketches or photographs of the changes, so you know, for instance, what's in the wall when you cover it up.

I think this should be part of every building code: No one can steal your sun, and all houses must have correct solar orientation. You have to think of the whole south side of a house as a solar collector.

Align the roof slope to your latitude, which should be the best angle for solar panels, either hot water or photovoltaic. It makes them a bit harder to work on, but it's easier to attach them firmly to the house, and they look like an intentional part of the design.

We tuned the house's orientation to our solar aperture, about ten degrees east of south. We run out of afternoon sun because of the trees and the ridge. We've had to top our closest trees to keep them out of the house's sun. And we built in thermal mass: seventy yards of concrete in the full basement foundation, and dark brickwork in the north greenhouse wall. I couldn't believe how much concrete we put in, or how much it cost, but you just have to do that part properly. We tuned the overhangs for summer shading—you can also use seasonal vegetation for shading, if summer heat is the problem it is here.

Try to eliminate pollution from inside the home. Vent the propane refrigerator and the water heater, because they burn and expel fossil fuels, for example. Limit formaldehyde-bearing materials, use nontoxic paints and sealants, and avoid CFCs. Even then, you need good ventilation.

Use local materials if you can, because they will be better and cheaper. We milled the substructure and rafters from a fir tree that was standing right where the house is. We found a standing dead fir that we thought might still have good wood in it; the mobile dimension mill produced six thousand board feet of beautiful lumber. The deck came from a redwood sinker dragged out of the creekbed; it's only medium grade wood, but it's perfect for decking, because it knows a lot about wet. I believe in sustained yield—you have to, if you live in a wood house—but we tried not to use endangered materials, and used recycled wood where we could. It takes longer, but preserves the forest.

Use the highest quality windows, the best you can buy. I had a problem with double glazed windows in the greenhouse, and my glass man informed me that's common in very hot applications. They turned white, then their seals failed. Be sure to get good advice. Use good doors and window framing, because the operating parts of your house should be the best possible.

Build air-tight, and insulate as intensively as you can, because it's easier to heat.

Linda: We used six-inch, kraft-backed fiberglass and six-mil vapor barrier in our six-inch outer walls.

Wes: Limit windows on the north side. You see, we have none.

Around here, the best way to heat, after solar, is with wood and an air-tight, catalytic woodstove. I can clear enough from the land to keep going and hardly scratch the surface. I figure if it rots it produces the same CO_2 as when I burn it. Woodstoves with small fireboxes and the high temperature required by a catalytic converter can be a problem, so research your stove carefully.

Plumb for solar hot water even if you can't afford the panels, and even if you just stub out the plumbing in the attic. Insulate all the runs, the hot side in particular.

Active solar hot water is more efficient than passive or thermosiphon. There are two problems to be dealt with: A, they freeze, and B, the controllers fail. Antifreeze in the collector and a heat exchanger tank takes care of A, and a system where the pump is matched to the sun with a panel takes care of B. We've got our solar hot water backed up twice, with a coil in the woodstove, and an instantaneous hot water heater.

Separate your grey water from your black water, and send it to artichokes or other crops that tolerate soap and grease. You can reclaim the heat in the grey water by running coils through your greenhouse beds. Low-flush toilets are a must; you want to keep your septic system happy. If it's not drowned with too much water, and stays uncontaminated from soap, it makes a better class of compost .

I prefer a 24-volt system. It cuts the amperage you're moving in half. Seek professional help when you're contemplating a full-blown system . . . the same as getting an architect's help, you'll save on mistakes and false starts. Talk to people that have systems equal in complexity to the system you want.

Linda: You will be amazed at what you learn from others.

Wes: Provide for future expansion. You can have it all, except heating, unless you have BIG hydro.

I no longer believe you need to double-wire your house. Direct current-to-alternating current inverters have gotten so reliable and efficient, and the quality of compact fluorescent light is so high—expensive, but once installed, people love getting 75 watts of light for 20 watts of power. Wiring for conventional 120 AC with 12-2 romex is much cheaper than wiring for both AC and DC. In this house, the smallest 12-volt wire is eight gauge, for lighting; that's too much copper in the wall. Instead, wire special circuits for 12-volt DC devices like pumps that are much more efficient than AC. You'll save a lot of energy if you plan carefully for zone lighting, and suit the fixture and source to the activity. And put in lots of extra outlets; you still never have enough.

Electromagnetic Fields (EMF) is something we should consider, even if we don't understand what it really means. We know there must be something to it, because the CIA was bombarding the Russian Consulate with it in the sixties. Wires in the wall don't worry me, but I want equipment at arm's length. Inverters should be outside the building with the batteries.

Put in extra conduit runs between the basement, crawlspace, utility room, and attic. You can never plan for the future because of new technologies and new interests. For me it was a radiotelephone, a new rain gauge. Photovoltaic modules are expensive, so you usually won't start with enough, but plan and wire for a huge array. Spend money on the wire. Our system, which has thirty-four panels and a micro-hydro plant now, recovers quickly and our batteries are fully charged by noon, so we can use the surplus running the dishwasher and the washer in the afternoon for free.

Trackers make sense for pumping water in summer, when you get 50 percent more power by tracking. In winter, trackers only add 20 percent. You use more energy in the winter, because you're inside more and the days are shorter, so it's less costly to add 20 percent more panels.

Before buying appliances, check for water and energy use. European appliances are best; American manufacturers are just starting to get the idea that efficiency is a good idea.

One last piece of advice: Grow your own food. Organically.

CHAPTER 8

BUILDING THE HOME ENERGY MACHINE

WHILE SEEKING TO BUILD FROM THE OUTSIDE IN, WITH FULL AWARENESS OF the particularities of region, neighborhood, and site, we should also undertake the planning required to build from the inside out. What functions, precisely, do we require of our home? Each of us will answer differently, and a family's answers will inevitably change as the individuals change, as the children leave, as the builders age. Anyone who builds an innovative, appropriate shelter hopes to be free from preconception but perspicacious about small yet important elements. If we intend to leave behind so much of conventional architecture and still build a successful home, we must carry with us only truly essential luggage. Happiness is a good measure for necessity. For example, it does not make us particularly happy to know that we are expensively hoarding eighty gallons of hot water in our domestic hot water tank.

And what is the widest temperature range within which we can live healthily and happily? The answer to this crucial question, factored together with the meteorological realities of our chosen site, determines for most of us the largest expense in our energy budget: the cost of heating and cooling our space. We must identify and answer these and many other energy questions completely and flexibly before building starts if we hope to build a successful independent home.

In the past, houses have been planned by assuming indefinite continuation of present circumstances. Selection of heating and lighting strategies, for example, has invariably been based on current costs. Scarcities and pollution have proved that we must anticipate steeply rising energy costs. We have also seen, in earlier chapters, that we should plan for cost-effectiveness based on original equipment cost *plus* the equipment's energy consumption over its whole life. Consider glass, which easily outlasts most houses; will high-tech glass, undeniably more expensive than single-pane glass but much more efficient, pay for itself over a span of time that will probably be longer than our own lifetimes? If so, then the better glass is the correct choice even if we shall not be around at payback time.

The primary energy consumers in a conventional house are space heating and cooling, refrigeration, hot water, and lighting. By conscientiously examining our largest energy uses, we will find ways to lighten our load and simplify our lives. Some of these ways of lightening the load should be taken into consideration before a final decision is made about the site, while other solutions must be found and incorporated into our house plans as we begin the process of building.

Naturally, a home's energy budget depends on region. In a tropical climate like Hawaii's, a well-designed home on a benign site will require neither heating nor cooling. In northern Alaska, even the best sited and best constructed home will carry a heavy burden in wintertime lighting and heating. Even if energy costs are tolerable at present, there are powerful reasons to seek to reduce energy consumption from the outset by eliminating loads or seeking alternative sources. This logic should be applied to each instance of energy use.

Paying the Price for Climate Control

Formal architecture might manage to create interior space that is visually connected with the outdoors yet completely isolated and regulated atmospherically. As soon as architects got the technology they needed to defy a site's natural climate, they defined a narrow comfort envelope within which, it was decreed, the human organism functions optimally: temperature between 68 and 72 degrees Fahrenheit (20 and 23 degrees Celsius), with between 60 and 70 percent relative humidity. Elaborate devices, strategic materials, and awesome energy are poured into achieving this goal.

Although other cultures abide by temperature extremes quite gracefully, one might almost expect this revision to be offered for inclusion in our U.S. Bill of Rights: "the right to life, liberty, comfortable interior

ENERGY DOMAIN	CONVENTIONAL SOURCE	REPLACEMENT
Heating	electricity	solar exposure superinsulation widened comfort envelope layered clothing wood: catalytic chimney excess hydro-electricity
Air conditioning	electricity	widened comfort envelope improved ventilation super insulation architectural measures evaporative cooling
Water heating	electricity propane/natural gas	solar solar assisted wood
Refrigeration	electricity propane/natural gas kerosene ice	super-efficient spring house shady side outdoor ice
Cooking	electricity propane/natural gas wood	raw foods solar cookers propane/natural gas
Lighting	electricity kerosene sunlight	natural light high-efficiency electric task electric
Appliances	electric	hand powered super-efficient solar-charged
Landscaping	fossil fuel	hand tools indigenous planting
Transportation	fossil fuel	human powered electricity shared (public)

Figure 29. In almost every energy domain, we can expect to find a partial replacement for at least part of our conventional energy usage.

space within a narrow range of parameters, and happiness of pursuit." Curious climatic abuses result from this fetish of ours—I have been too hot in Boston's winter, and too cold in Fort Worth's summer—as do a variety of ills, from Legionnaires' Disease to Sick Building Syndrome, which result from our precipitous willingness to defeat the environment with biologically disruptive technologies. Like willful children, as soon as we found a way to defy nature, we did so, without ever contemplating the repercussions. When the Arabian wake-up call came in 1973, and energy awareness gradually took hold as a result of the oil embargo, the supposedly sacrosanct comfort ranges were redefined, and a be-cardiganed President begged us to comply.

Our reliance on brute force, high-tech heating, ventilation, and air conditioning (HVAC) solutions is largely the result of our abandonment of the homestead for many hours of every working day. When we stay at home, it is easy to open and close windows, feed the fire, and otherwise regulate our home, but on most days, when we return home exhausted from a hard day's work, we expect to be greeted by a welcoming, comfortable, well-regulated environment. There is a sharp irony to the fact that many of us must leave our homes to earn enough to support the machines that keep our homes habitable in our absence. Who, precisely, are we working for? The energy we buy in order to work—fuel for the commute, and the cost of maintaining a comfortable home space even when no one is home—is a hidden tax on our productivity which many independent homesteaders refuse to pay.

Some of my heroes, vigorous old ones who create beauty even as they age beyond my youthful imagining, showed me as I planned my house that a life connected with nature's grittiness is longer and more vivid than one lived within the comfort envelope. Now, a decade or two later, I willingly take part in the spirit of the season by wearing several layers, or as little as possible, as the climate requires, in preference to expensively denying the elements. As already noted, my cat's fur adapts to the season whether the house is within the envelope or not; in matters like this, my cat is seldom wrong. To put it quite simply, I enjoy, as an independent power producer, living in space that buffers me within rather than isolating me from the weather, and invites me to participate in the seasonal responses of my local ecosystem.

Even before widening our tolerance for temperature extremes, we may find that sensible design will enclose space that regulates itself quite readily. Two key factors apply: incident solar radiation (which yields a coined word, *insolation*) and thermal mass. Using glass, we can invite sunlight in to heat our space to a livable temperature. Sufficient thermal mass (material that holds heat well, like masonry, earth, or water) within our space will attain a comfortable temperature, then act as a buffer, radiating heat or coolness to keep the space comfortable. Such self-regulation, with no ongoing expense, is a grand solution, but attaining it is not at all simple. Too much glass, and the temperature varies wildly no matter how much mass is enclosed; too much mass, and the temperature never becomes comfortable. The ideal glass-to-mass ratio is delicate, site- and design-specific, and therefore hard to plan for, but instead must be found through experimentation. In a breakthrough house, one unlike any house built before, this can be complicated, because glass and mass are often costly to include and difficult to adjust. Most architects find it easier to avoid the issue altogether by employing powerful HVAC units. One possible solution is to start with a small, conceptual space, and tune it until the principles are grasped, then enlarge along the same proportions. This is unworkable for architects, who wish to present completed totality, but fits the alternative builder's gradualist mode exactly. Another tactic is to provide a means to isolate exterior glass and mass from the rest of the house, as Wes and Linda have done by interposing operable windows, so that the active thermal components may be opened to the house when they are working, and closed when they are not.

Wood and glass houses, though quickest to build, have almost no thermal mass and are therefore at the mercy of ambient outside temperatures. Concrete, rock, masonry, tile, earth, water, and other dense materials provide thermal mass. If placed where they intercept the sun and are directly heated, the arrangement is said to have *direct gain.* Gain can be improved by making the mass a dark color, so it absorbs the incident sunlight maximally. Where sunlight strikes materials such as wood, cork or vinyl tile, cloth, or other less dense materials that hold heat poorly, and the heat is given up to the air which then heats available thermal mass by convection, the process is less efficient, and is termed *indirect gain.* Floorplans where the southern quarter is dedicated to direct gain thermal mass are most successful; this quarter is the right place for a greenhouse.

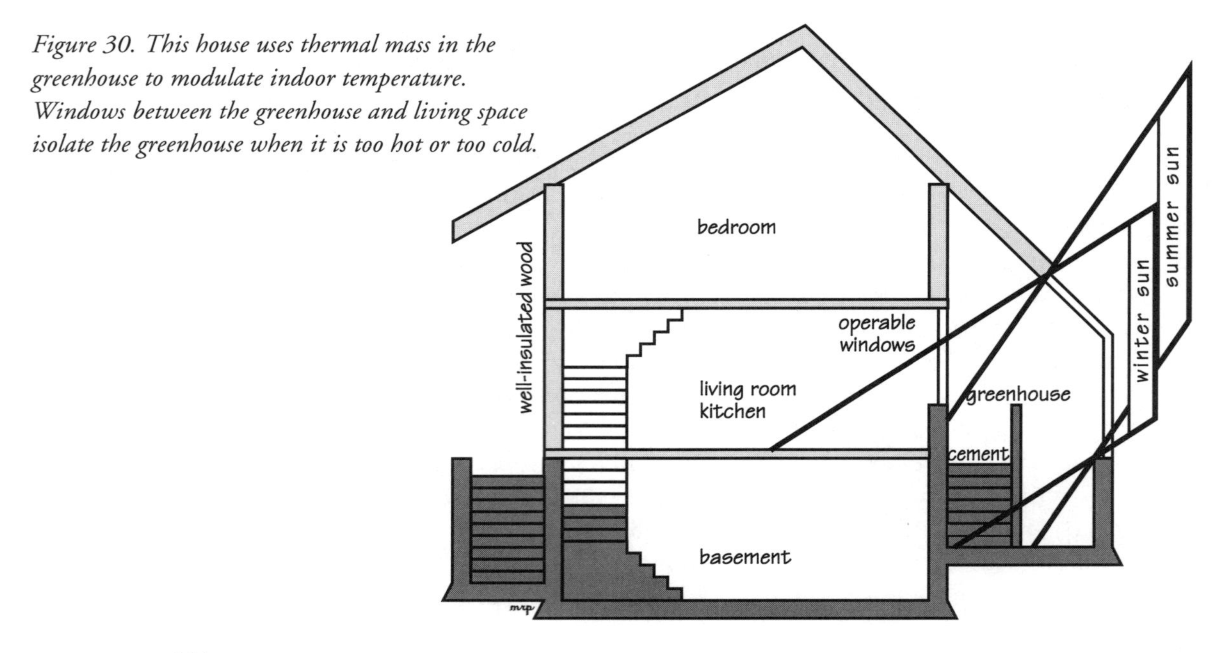

Figure 30. This house uses thermal mass in the greenhouse to modulate indoor temperature. Windows between the greenhouse and living space isolate the greenhouse when it is too hot or too cold.

Pounding tires: Ken Anderson's story

Ken Anderson is an architect who works for Michael Reynolds. He is building his own earthship at Star, one of two earthship communities near Taos, in northern New Mexico. His enthusiasm and delight in this approach to homebuilding kept us both smiling as he talked. (Michael Reynolds tells his own story in chap. 14.)

After helping others build their earthship homes, Ken Anderson starts work on the first tire for his own home at Star, near Taos (photo courtesy Pamela Freund).

I came out here from New Jersey, Philadelphia, and New Haven, and there's a good chance I'll stay here until I die. My goal, when I graduated, was to get as far away, philosophically and geographically, as I could. I had no idea how to build the buildings I was drawing.

When I got here, Mike Reynolds put me to work on an earthship, and I stayed with it from scratch, a year and a half, from the first hellish tires. I really learned architecture on that job. I learned everything I know about building that way, and right now, I can't say I remember much I learned at Yale. A love for architecture, I suppose, and an open mind about design.

Now I've learned to draw buildings the way I'd build them: first the hole, then the tires, the greenhouse, the plaster fill. When you're involved in building, you understand the whole system. It's unlike basements, which nobody understands; we know they're down there, and have things in them, but they're pretty mysterious, really.

The idea of the first *Earthship* book was to make it possible to build an earthship by yourself. Some of the best earthships are owner-built from the first book. "Who needs to draw it? Let's just go out and build it." That worked for some, but in general we learned from people out there that there was too much bureaucracy, so we made a set of generic plans that give the inspectors, building departments, and engineers everything they need to see that we've thought this idea through. For fifteen hundred dollars you get the generic plan set and enough consulting time to get started; you build the generic structure, and get signed off, then make custom modifications with your finishes. We believe the generic earthship is a very sound and satisfactory home.

There are well over a hundred tire houses scattered around: New Mexico, of course, and Arizona, California, Oregon, Oklahoma, Washington, Colorado, Kentucky, Tennessee, Arkansas. The oldest ones are eighteen or twenty years old, and still look great. Mike is always improving things, seeking new solutions to problems that come up—you saw the hay-bale temporary house? We've completely cleaned Taos of tires, so last year we had a tractor trailer load brought in from Colorado.

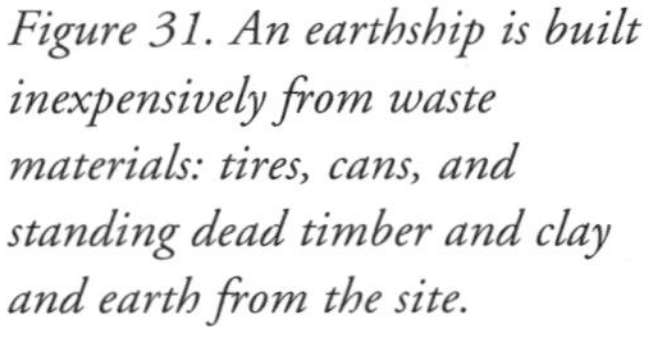

Figure 31. An earthship is built inexpensively from waste materials: tires, cans, and standing dead timber and clay and earth from the site.

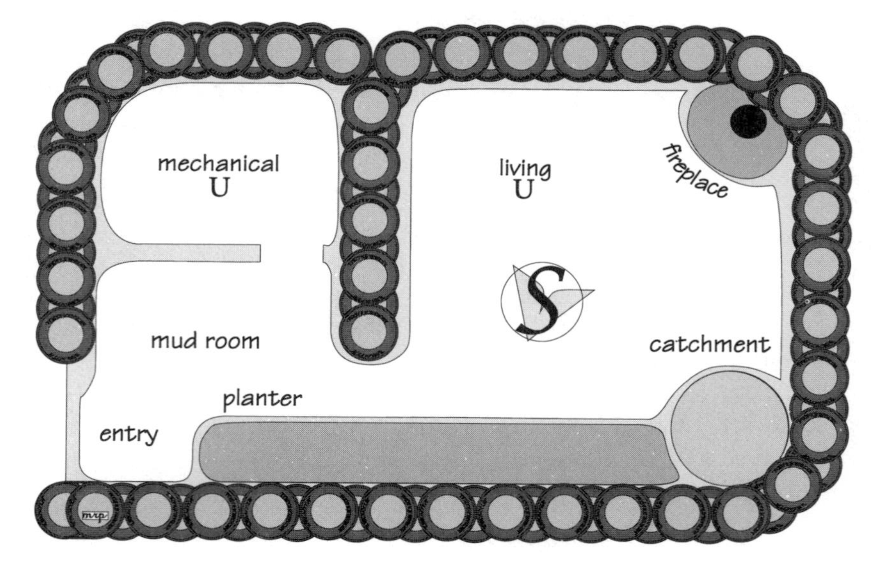

Every earthship we've worked on has a building permit. We have made a conscious effort to establish the viability of this building method, and to work with building departments so they would know what we were trying to do.

It takes five or six hundred tires to make a small house; if you start with the house excavated from a hill, you can cut down on the number of tires. There are about three wheelbarrows of earth in each tire, and it takes two people about fifteen minutes to pound a tire. Maybe you can do thirty tires a day, and at the end of a few days, you see the house forming up, and you get a second wind. As the tires fill with tamped earth, they swell and interlock with the row below, so the walls are very strong. The walls are usually U-shaped (and we call each bay a "U") which also adds strength. We usually alternate a full U, sixteen feet wide and eighteen feet deep, used for living space, with a mechanical U, which is shallower and usually houses battery and electrical systems, water pumping and filtering, and other utilities. The simplest earthship is a full U and a mechanical U.

Along the top of a tire wall, you put in a cement and steel or a wood bond-beam, to hold the wall together, and for attaching the roof. The sloping front wall involves some pretty special construction techniques, and the glass isn't cheap, so generally you need a good carpenter to do that part.

The traditional way of adobe roofbuilding, using vegas and latillas, works very well, but we're working with trusses, too, which use much less wood. You can buy vegas at the lumberyard for three or three-and-a-

Figure 32. Dug into the ground, the earthship's thermal mass keeps it cool in summer and warm in winter.

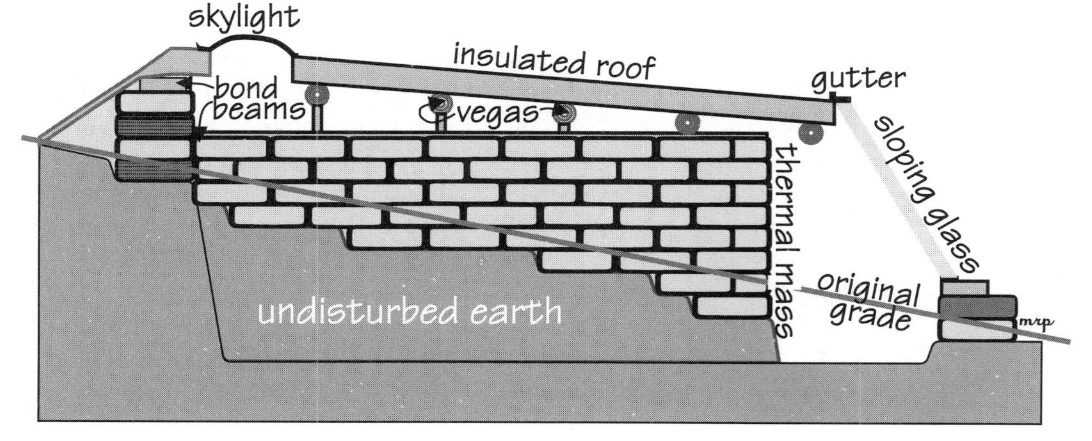

half dollars a lineal foot, which is expensive, or harvest them in the national forest, mostly standing dead, with a vega permit which costs about three dollars for each vega, then strip them, which is labor intensive. We use some fairly high-tech sealants, six-mil plastic sheeting to seal over the tires and berm, and then a modified bitumen roof coated with acrylic over the styrofoam roof insulation.

Earth rammed into the elastic tires of an earthship make an excellent wall. It may take six hundred tires to make a small house.

I have chosen my building plot at Star, and I've been stockpiling tires and cans in a storage bin. The groundbreaking is this weekend! I expect to get the shell—walls and roof—put up this summer, so the 'ship can start charging up its walls. The tires are like batteries, and they take up to a year to get charged with heat. By winter, it will stay a constant sixty-five to seventy. Next summer, we'll finish the interior and install the electric and water systems, and it will be ready to inhabit.

Mike's idea about the earthship is that it should be self-sufficient. Electricity comes from eight 51-watt modules on the roof and ten golf cart batteries. The electrical center makes the unusual connections, and gives the electrician a standard breaker box that he knows how to work with. We do the same thing with the water system; we want to give the tradesmen systems they understand, with standard fittings and connections for them to tie into. All the lights are 12-volt DC, compact fluorescent, and all plugs and appliances are 110-volt AC. Water comes from the roof, and is kept in a catchment inside the 'ship. We figure that you can live with eight inches a year of rain if you recycle the grey water into the indoor planters and use a solar or composting toilet. The water system consists of a simple system of filters and a pump, and runs off the electrical system. These two systems cost between sixty-five hundred and eight thousand dollars, depending on what you need, although we've done much more complicated systems.

Star is about forty-five minutes from Taos by four-wheel-drive (although it will take less time once Star is more fully inhabited and the roads are improved). It's near the Rio Grande gorge, and the views are great. You buy the land under your house and membership in Star for two to ten dollars a square foot, depending on whether you choose the high-density or low-density area. High-density means a minimum of fifty feet between houses, and low- means four hundred feet. I really haven't decided which I like better, but the land will be affordable either way. The expensive parts, not counting the labor (and about 85 percent of earthships are owner-built) are backhoe excavation, tire and can procurement, metal lathing, and the glass, cement, and lumber in the bond-beams, roof, and glass wall. My earthship itself won't cost thirty dollars a square foot, so for under fifty thousand, and two years of work, I'll have a nice home that completely takes care of me. Within a few years, we will establish an architecture office at Star. I'll probably cut down to working three or four days a week, providing architectural services and construction management to my future neighbors.

Simple, Passive, and Massive

In a thermal-mass heated home, it may take a year or two for interior temperatures to "pop" or come to equilibrium, and even then, a certain amount of heating or cooling may be required to deal with seasonal excesses. In northern New Mexico, where adobe houses are the regional norm and where the earthship was invented, adequate ventilation must be provided or interior temperatures will exceed the outside where breezes blow. In winter, a small amount of heat will be required, or the dwelling's temperature will drop into the high 50s during stormy periods when it goes down to -10 degrees outside. In the same area, wood-frame houses without heating and cooling are intolerable most of the year.

Massive stone or brick fireplaces, concrete and masonry furnaces, and other self-heating structures can also provide thermal mass. If the combustion techniques used to heat such a mass are clean and rationally fueled, they work exceedingly well. The siberian stove, for example, need only be fired once a day even in a harsh climate; its recirculation of exhaust and its high-temperature firebox assures that the smoke that finally escapes has been thoroughly cleaned of noxious combustibles, and the substantial pile, once heated, continues to radiate heat for hours. High-tech schemes for using electricity to actively circulate warm air through rock bins and dimorphic salts play well in popularized science magazines and attract the attention of government boondogglers, but have proved to be too complicated to work well. In space heating and cooling, simple, passive, and massive are best.

High Ceilings, High Heating Bills

Heat rises, and in the absence of active circulation the top eighteen inches of a heated room may have a temperature ten degrees above the eighteen inches just over the floor. Air currents set up when people move about help mix this precious warm air only if the distance above their heads is limited. All this argues in favor of low ceilings. We have been conditioned to believe that anything less than a foot above our heads feels very close (although I have been in experimental homes with very low ceilings and found myself adapting quickly). But how many of us are taller than six-and-a-half feet? Eight-foot ceilings, we are told, feel dignified, and are convenient to build because panelling usually comes eight feet tall, but eight-foot ceilings have nothing other than custom to recommend them. Building codes generally permit ceilings as low as seven feet six inches.

Considering that approximately 10 percent of our heating budget goes into heating an additional six inches, I hope we can agree that seven feet six inches is ample.

While on the subject of heating, firewood management is a worth a bit of attention. When home systems are planned starting with the energy to be used, we will be able to think about ways to make heated air circulate naturally through the home. But we must also plan how firewood gets from bulk storage to the woodstove. We try, in Caspar, to get in a week's supply before a storm breaks; special firewood doors make this an easy operation.

Doors Indoors?

Doors inside a small house make sense only when we wish to heat a single room. Doors represent a compromise between conflicting needs: we want be able to move from room to room, to foster or prevent air exchange, and to interpose a temporary wall across an opening, for privacy. Doors should therefore be light, which means small. Conventional hinged doors always waste the floor space and wall space within and behind their arc, and closed, they interfere with air circulation. Archways work better in almost every way. Privacy can be created by juxtaposing walls, screens, and archways so that rooms have the sense of privacy desired. As with any radical idea, it is best to prepare well in advance for an easy retreat; use standard doorway sizing and framing practices if you have any doubts about adopting my suggestion of eschewing doors.

Insulation Is Everything

Thermal mass heating, and any other strategy that relies upon separating indoor and outdoor temperatures, calls for insulation measures in direct proportion to the difference between outdoor seasonal ambients and the indoor ideal. Considered over the life of a house, it is easy to conclude that the thickest insulation, the highest R-value, the best glass, the most thorough reduction of infiltration, is cost-effective from the vantage point of energy savings. Good siting, excellent design, and superb insulation should make it possible to build a passively heated house in almost every habitable climate. In other words, most energy expended on space heating and cooling is a direct result of bad planning, negligent design, and sloppy insulation. Now, as the energy buzzards come home to roost, we pay the price.

Changing technology may offer a chance to improve on earlier solutions, and new houses will be built to accommodate such changes. Low-emissivity, high-tech glass was not available in California when I put windows in my house, nor was it in my budget based on short-sighted, pre-1973 energy cost-benefit calculations. I was wrong. Two decades later, the cost-effectiveness curves have crossed: energy costs are stubbornly climbing, and the cost of high-tech glass is lower than ever due to its enthusiastic acceptance by the building fraternity. Replacing single pane glass with low-emissivity windows will probably eliminate my need for a woodstove on all but the coldest winter nights.

The standard way to measure the heating and cooling requirements of a house are *degree-days*, the difference in degrees between maximum (for cooling) or minimum (for heating) temperatures and the comfort baselines, totalled over the average year. Taking 80 degrees as the high comfort baseline, for example, if the outside temperature climbs to 100 on a given day, the house requires cooling capacity of twenty degree-days. If this condition persists over a one-hundred-day cooling season, the house will require sufficient cooling for two thousand degree-days a year.

Good insulation and adequate thermal mass should be able to buffer, in all but the most unforgiving climates, nearly all of our heating and cooling needs. Problems result where ambient temperatures remain far above or below the comfort baseline, day and night, for long periods. In the desert example given above, where daytime temperatures climb to 100 degrees, everything depends on nighttime; if outside temperatures drop to 70, then thermal mass may be recharged by opening windows, and during the daytime, insulation (including shaded, southern glass areas) and thermal mass should keep the house comfortable. If nighttime temperatures remain high, usually due to lack of wind currents and large thermal masses in the environment, we grudgingly grant that a problem exists, which may justify expending energy.

In high degree-day cooling locations, intelligent use of overhangs and seasonal shading are crucial, as is sensitive siting, to take advantage of any shade and to capture any breeze. Indigenous structures built in such unforgiving sites offer a number of ingenious passive measures for surviving the withering heat. Often, not far below the blistering surface, the earth maintains a constant and relatively low temperature, so underground houses work well. Cooling towers, which use convection to circulate cool air from underground or from protected sources or which provide evaporative cooling, and wind scoops positioned to catch cooler breezes a dozen feet above the ground, cost nothing to operate after they are constructed. All else failing, the sun itself may be converted, using photovoltaic panels, into energy to run cooling equipment. The worst solution of all (yet the most applied) is to build a poorly conceived and uninsulated house, and then importing at great expense to occupant and planet the electricity required to run a powerful air conditioner.

Figure 33. In desert areas, a cooler breeze often passes above the rooftops. These wind scoops, from Egypt, Peru, Pakistan, and Afghanistan, direct this precious breeze through the living quarters. From Commonsense Architecture: A Cross-Cultural Survey of Practical Design Principles *by John Taylor (W.W. Norton). Copyright 1983 by John S. Taylor. Used with permission of the author.*

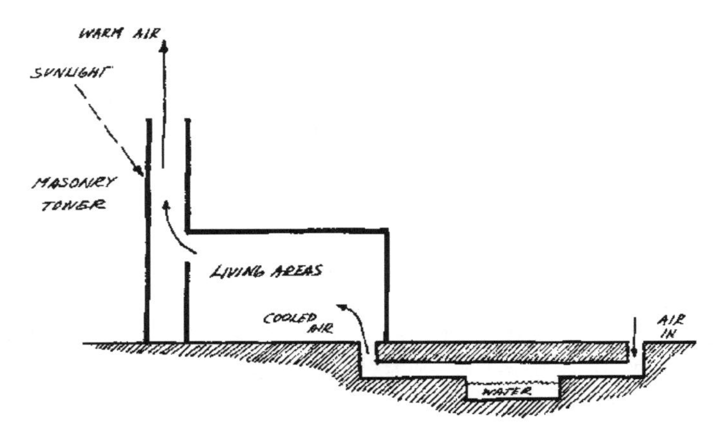

Figure 34. A Middle Eastern cooling tower is heated by the sun, and the air within rises, drawing cool air behind it. If the air is drawn over water, it is cooler and moister (From Commonsense Architecture *by John S. Taylor. Copyright 1983. Used with permission of the author.)*

Insulation and proper siting are also the best—and practically the only—passive remedies in high degree-day heating locations. In arctic, mountainous, and far northern lands the sun is too feeble, and too often obscured, to offer much opportunity for collecting the sun's heat. The earth may be deeply frozen and therefore not much use in providing protective buffering. Streams may also be frozen solid, and the only possible source of excess energy may be the wind. The indigenous response is to use the snow itself as an insulator, to dress warmly, and to hope for spring.

In less extreme cold locations, like the Colorado Rockies, where degree-day heating requirements are nevertheless high, the solar/thermal-mass solution often works well because the sun still shines brightly, despite lingering cold from wind and snow. Running water offers hydroelectric potential, which is the only alternative energy source that can be reliably and economically devoted to supplying electricity for heat. And in most such places, forests are abundantly regenerative, and can be counted on to provide a fair quantity of winter fuel from a reasonably modest area of woodland.

Our preference for homes surrounded by sufficient land takes us back to the time before the dependent home, when it was the job of the sun, the woodlot, and the prevailing breezes to condition interior space. Heating and cooling are examples of tasks not well suited to electricity (unless we have a surplus). Before electricity, humans found it difficult to preserve a high quality of life in an urban settlement of a quarter-million people or more, primarily because that many people consuming energy in close quarters were dirty and dangerous. Now that electricity allows us to do the dirty work elsewhere and deliver energy in an apparently clean and extremely concentrated way, our cities are able to support enormous masses of denatured humanity.

Building in Harmony with Sources

By taking our energy needs into account very early in the planning stages of a new, independent house, it should be possible to integrate a variety of sources and vastly improve our homestead's performance.

The solar energy which falls on a south-facing roof can be either a curse (as an undesirable heat source for the space within) or a blessing (as a source of electricity and hot water). We know that low-voltage electricity does not travel long distances efficiently. Moreover, we find that the area required for solar panels—photovoltaic modules plus solar hot water panels—often corresponds nicely to half of the house's roof area, but only if the roof faces south. A very satisfactory and cost-effective accommodation can be reached by dedicating roof space to capturing the sun, thereby incorporating the solar panels into the house's existing profile. One of the more progressive utility companies, Sacramento Municipal Utility District (SMUD), is petitioning their ratepayers to "Give Us Your Roofs" for precisely this purpose (see page 32).

To take advantage of this design economy, the house's roof must be planned to align with the site's solar aperture, where the sun may be captured most effectively. The sun's path across the sky has two variables, time of day and time of year. At noon on the summer solstice, the sun will ride at its highest point, due south, at an angle above the true horizon of the latitude plus 23 degrees. At noon on the winter solstice, it will appear due south at an angle of the site latitude less 23 degrees. (The earth's axis is angled at 23 degrees to the sun's plane, which causes seasons.) At noon on the equinoxes, spring and autumn, the sun can be found due south at the angle of the site's latitude above the horizon. Through the rest of the year, the sun commutes between these points. The proper angle of exposure for any solar capture device corresponds to the site's latitude plus or minus the seasonal fraction of the sun's semiannual range of 23 degrees. A fair compromise, choosing the site's latitude as the roof's angle to the horizontal, only varies from the optimum by approximately 23 degrees at the solstices. If maximum electrical loads are expected in the dead of winter, as is typical, the savings realized by mounting modules on the roof (as opposed to free-standing racks) can be invested in enough additional modules to compensate for the inefficiency of slightly misaligned modules for part of the year.

Please note that compasses seldom point to true north, but rather to magnetic north. Either get the local correction before using a compass to orient the house, or make careful observations using the time-honored Boy Scout method: with a tall stick and pegs mark the shadow's end over several days; the shortest shadow points due south.

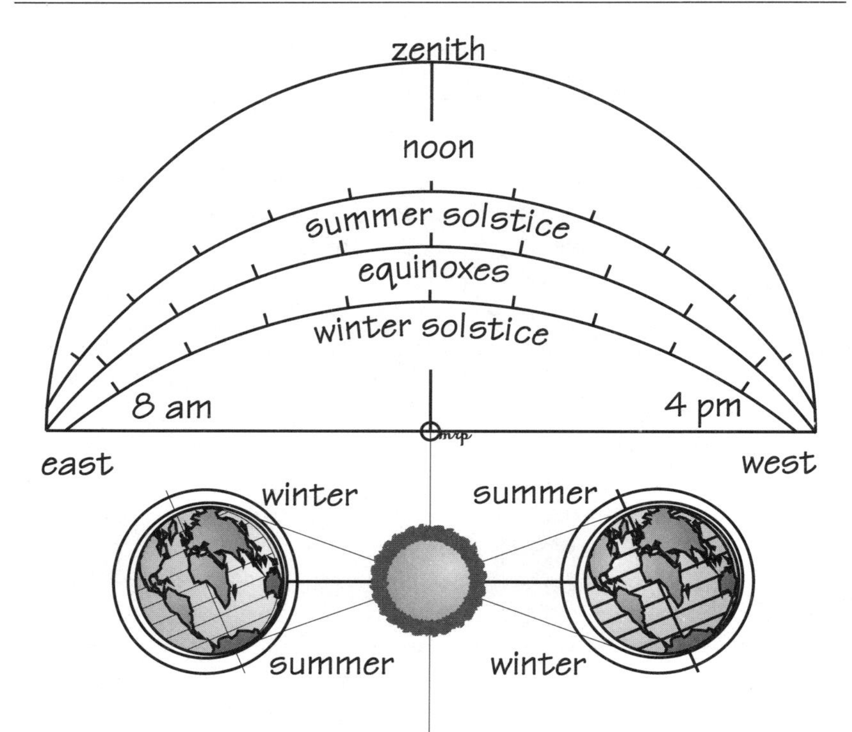

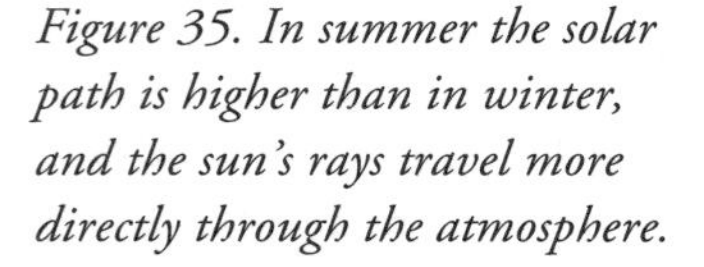
Figure 35. In summer the solar path is higher than in winter, and the sun's rays travel more directly through the atmosphere.

Carpenters express roof pitch as the ratio of rise to run, where run is always twelve. In Caspar, where winter sun is at a premium, I have devoted a section of roof to solar harvesting, and angled it to be perpendicular to the sun about a month before and after the winter solstice, or 18 degrees higher than my latitude, a precipitous pitch of fourteen in twelve. My best solar aperture is slightly west of due south because of trees I do not care to cut and typical wintertime cloud patterns. Solar energy experts have special tools for predicting the solar aperture and its changes over the solar year, and can help get the roof pointed exactly right. Getting it wrong can easily subtract 20 percent efficiency from your array.

Coincidentally, increasing the roof pitch with latitude serves the need to employ steeper roofs in snowier lands. Since snow will cling to a glass surface at or below a 50-degree pitch (particularly if the surface is dirty), some thought must be given, if the homestead relies on PV for snow season power, to convenient ways to clear the snow.

Roofs can be used to capture another precious commodity, water. In much of the world, roof catchment is the primary source of domestic water, but in conventional American homebuilding, roof run-off is considered more a nuisance than a resource. Roof-captured water requires careful treatment, particularly in a suburban or urban setting, but as our water resources come closer to their limits, we ought to put these pennies from heaven to good use. Thanks to gravity, it takes very little effort to divert roof run-off to a holding tank; it takes even less when this intent is incorporated into your plan early on. If the run-off is to be used for drinking or to water edible plants, care should be taken to select a roof surface that does not spoil the water with leachates or unfilterable impurities. The manufacturers of the roofing material will be able to provide reliable information.

Someday in the near future we will see photovoltaic roofing materials, and the economy of renewables will be complete. What would shed precipitation better than a glass roof? Until then problems with expansion and module maintenance make it necessary to install the panels on top of a roof covering with a life expectancy at least as long as that of the modules, which we know to be approximately thirty years.

Photovoltaic panels are very sensitive to heat: they don't like it, and become less productive as they warm up. This perverse behavior cannot be corrected by the usual means—you do not want to shade your array—so the next best thing you can do is attach them to the roof in a way that keeps them as cool as possible. My modules are firmly affixed to vertical rails that provide a chimney which draws cool air in at the bottom. Without this chimney, the temperature behind my panels will be as much as 30 degrees above the ambient on a sunny day, which decreases module performance by about 10 percent. Panels don't care how they are aligned, but the chimney effect can be enhanced if they are installed with their long dimension oriented vertically.

You may note that I emphasize firm attachment of array to structure. When strong winds blow, modules are expensive, fragile, sharp-edged kites waiting to fly. Aficionados of the electrical code don't worry as much about this as I do; they worry instead about another unlikely but potentially devastating event, lightning. And they want every module and metal part individually grounded.

There is a brisk argument in the solar community about the efficacy of racking and tracking as compared to solid installation. By seasonally tinkering with the horizon angle of the modules, small gains in efficiency can be made; by adjusting module orientation for time of day, relatively greater gains can be made. Racks and trackers are usually rated to withstand 120-mile-per-hour winds; in Caspar, or wherever the weather gets serious, that is inadequate. As a veteran of winter's lusty gusts, and having seen the effects of serious storms and weathering, I hold with those who favor solidity.

To my eye, a tracker full of photovoltaic modules in the yard is just slightly less offensive than a powerline swooping in from the street: it bespeaks a lack of forethought and a hint of impermanence. Many who look to home energy harvesting as a long-term power source are attracted to photovoltaics because of their elegant passivity. If it is impossible to orient the home's roof correctly, it would be better to see panels on an outbuilding with an appropriately sloped roof than a tracker in the yard. Possibly the batteries and power-conditioning equipment could live inside the same shed.

To maximize summertime energy harvesting (for agricultural pumping, for example) trackers make sense in a benign climate. Active trackers which use electronics to follow the sun are accurate to within half a degree, and get on the job first thing in the morning, while passive trackers that use freon (a CFC, but safely encapsulated, we hope, and not prone to escape) are less accurate, and take half an hour to wake up in the morning. Both are beneficent technology, but the active trackers are better, and of course cost twice as much. A tracker might improve summertime energy yields by 40 percent, and wintertime yields by half that, and usually costs about half as much as the panels it carries. In other words, it is more cost-effective to solidly mount 20 percent more modules to maximize the winter energy harvest. Over twenty years, I believe a strategy of buying 20 to 40 percent more panels and mounting them solidly on the roof will prove to be the most trouble-free and economical. New technology arriving in the marketplace in the coming year may change my mind, but presently available panels with optical concentrators must be kept pointed directly (to within a quarter of a degree) at the sun, and existing active trackers are barely capable of pointing this precisely.

The *solar fraction* is the percentage of a household's energy requirements that can reasonably be expected to be generated by harvesting solar energy. In some sites, where cloud cover is rare and the sun beats down unimpeded, the fraction is 100 percent. An old government map of solar resources showed such spots in red, while less gifted sites ranged through the spectrum to blue where, at one time, it was thought that no solar capture was feasible. Vermont was all blue, but this estimate, like the report of Mark Twain's demise, was premature. Richard Gottlieb, perennial candidate for governor of Vermont, has been cheerfully harvesting solar electricity for almost two decades, and reports that in his part of the state (the south) the solar fraction could be as much as 70 percent. Deficits, of course, may be made up with more modules and a bigger battery bank, but it is more sensible to develop a second energy harvesting method if possible.

Alternative Energy Gotcha!s

Things that spin, sing. Where photovoltaic panels rest in the sun, quietly generating electricity, generators that convert rotational energy into electricity—wind-spinners, hydro-turbines, and motor generators—make noises that may be disconcerting and that certainly intrude on rural peace and quiet. Owners of homes with micro-hydro turbines and wind machines close by told me that they found the sound reassuring in the night, but I wonder if this sentiment is shared by other family members and neighbors. These devices are best at a distance—wind machines, in particular, as far away as possible—and distance creates problems. As noted above, electricity loses energy when it travels. The grid that connects all the plugs in America wastes during transmission up to half the electricity it carries. As we saw in chapter 4, distance costs power, and so does conversion. Where possible, we want to generate the same form of electricity that we intend to use. Low-voltage electricity loses more energy, proportionally, and so spinners and turbines at a distance usually spin high-voltage alternators or generators. Furthermore, moving parts, particularly those in contact with hostile forces like water, wind, and heat, inevitably require periodic attention and replacement. The extra maintenance requirements, and the conflict between electricity's preference for proximity and our need to keep spinners at a distance, often mean that PVs, which look to be more expensive when only the price of another kind of generator is considered, are cost-competitive when the whole installation is accounted for. Add the fact that PVs are likely to decrease in cost and increase in efficiency more than competing technologies, and you have a compelling argument in favor of building a good site for PVs even if you don't expect to use that source right away.

When we can, we build our houses out of the wind. Wind strips heat out of a house, and wind-chill makes a place very uncomfortable, so our homesites are seldom good windsites. Only on the flattest of plots will homesite and windsite be the same. On the other hand, since we seldom build houses where the wind resource is best, because it would be too windy there, it is easy to distance ourselves from the *thwocka-thwocka* of these noisy devices. Hydro turbines need less distance, as their high-pitched whirr is easily kept inside the massive cement housings they prefer.

Just remember: the wind machine is best not attached to the house, and the micro-hydroelectric turbine somewhere other than under the bed.

Planning for Maintenance and Change

Plan to install extra electrical outlets. There never seem to be enough, and electricians report this to be their commonest and least favorite electrical job. Install twice as many as required by code (which usually calls for one every twelve feet) with special attention to areas where you expect to work or concentrate electrical equipment. Underwriters Laboratories reports that many electrical fires result from faulty or damaged extension cords. This danger can be circumvented by anticipating where extra outlets will be needed, and wiring accordingly. Trade electricians usually charge by

Mobile Homes

From an energy standpoint, it is hard to imagine a more miserable shelter than a mobile home. Using aluminum for lightness (a material that embodies enormous energy because it is refined in electric furnaces from bauxite, and is further cursed with good heat conduction properties), these thin-walled, barely insulated, single-glazed monstrosities are serious warts on the face of our nation. They are independent in the sense that their occupants can, at a whim, pump up the tires, kick out the blocks, and be gone, but they are horribly wasteful. In fact, their impermanence commits the occupants to a life of unwitting disconnection from the land their mobile hovels infest.

To heat or cool a trailer requires roughly triple the energy used by a properly insulated house of equal size, and trailer residents, who invariably pay the energy bills, are condemned to perpetual slavery to the energy mongers because of this inefficiency. In a trying environment, it is practically impossible to operate anything but the smallest mobile home on an alternative energy system, because it takes too much energy to keep them warm or cool. Furthermore, they are firetraps and they are containers for a horrifying collection of toxic vapors.

Evaluated by the freedoms enumerated in chapter 4 (rational construction costs with sustainable materials, reasonable expenses to operate and maintain, and a responsible way to recycle and abate the structure at the end of its useful life), the mobile home is a mistake from beginning to end. Significantly, elsewhere in the world, not even the most disadvantaged people consider them remotely habitable.

Champions of the mobile home offer the excuse that these may be the only kind of habitation affordable to low-income folks. That's absurd. So much goes into making the damn things mobile that equal energy and equal care put into a stable dwelling would provide a much more functional and beautiful home at much less cost to the environment. It's also important to ask—to what extent have the lax health and safety standards applied to trailers made their existence possible, while similar relaxations have not been applied to solidly constructed low-income housing?

The advice of many who had moved onto undeveloped land and lived in a trailer while building their home is, "Don't waste your time and money." Camp in a tent, and build a guest cabin to weather the first winter in, while you learn about your site first-hand.

the outlet box, not by the hour, and bid their jobs based on the box count; on my travels I heard prices as high as twenty dollars a box. Naturally, if our eye is on the job cost, we may be tempted to cut costs. As always, remember that you will pay for such false economies the whole time you live in the house. It is much cheaper to install an extra outlet when the walls are open than to add one afterward.

Make maintenance easy. If simple tasks like changing lightbulbs and water filters are inconvenient, we are likely to leave them undone. Proud owners showed me many innovative arrangements for making periodic maintenance easier, from built-in light fixtures with hinged translucent covers held closed by magnets, to built-to-size trays under the sink so that the welter of household chemicals kept there could be easily set aside while changing filters or working on plumbing. Traps and pipes inevitably drip when being serviced. By building cabinetry that easily allows us to slide in a plastic tub or tray before the dribbling begins, and out again without spilling, and by designing the plumbing so that connections are above the places where a tub can be set, we make the chore quite a bit more pleasant.

Placing lights in the dark places where someone must occasionally work is a novel idea that often occurs too late to homebuilders, even if they expect to work on their own systems. The same goes for outlets in places where power will be required only occasionally; by wiring with such foresight, our tool-gathering task is lessened by at least one extension cord.

CHAPTER 9

PIONEERING AND SETTLING THE NEW ENERGY FRONTIER

From the point of view of efficient use of energy, by far the most successful homes I visited were built in the last two decades from the ground up in accordance with a builder's vision. These homes seemed to have grown from their conceptual hearts outward, adding layers as a tree adds rings.

Dimetrodon, a wonderful twenty-year-old modern dinosaur of a manse near Warren, Vermont, is the collaborative work of a group of young architects, one of whom, William Maclay, still lives with his family in one of its small apartments. The primeval Dimetrodons, now called the solar dinosaurs, heated and cooled themselves by circulating blood through their sail-like backs. The residential Dimetrodon was envisioned by its builders as an experiment in shared housing, where small, personalized private living spaces could cuddle together around public space. As one of the original co-housing projects, Dimetrodon is a success. Technically, it exemplifies a pattern that I found in many of the first independent homes. To my eye, its windows seemed to be pointed the wrong way, to the north, away from the sun. William agreed—"Well, we had to make a few mistakes,"—and he then explained how the south side's active solar heating system was supposed to work. Water trickling over a blackened surface was heated by the sun, captured, and circulated through the building, providing primary space heating. The south-facing roof was dedicated to this promising, low-tech concept, which proved not to work very well in practice, and was abandoned. (Someday, perhaps, the south roof will be covered with sparkling, sapphire-blue photovoltaic panels and state-of-the-art solar hot water heaters, and the designers will be solar heroes and visionaries, their first setback forgotten.) Other aspects of Dimetrodon work well: the large, innovative, homemade back-up heating plant is efficient, and residents report enjoying its cooperative maintenance through the winter. Like a fire-breathing dragon, the heating plant dominates the basement, but its exposed pumps and pipes make it easy to understand, troubleshoot, and service. The small apartments are clustered together to conserve heat, and constitute a self-contained, supportive neighborhood. Fittingly, Dimetrodon was planned as a skeleton upon which residents could flesh out their spaces according to their personal visions, and so every space is different. Inside the limited confines, every cubic inch has been made to function, which has mothered a wealth of inventiveness.

Neighboring houses on Prickly Mountain, mostly architect-built, single-family residences, are correct, formal, and unimaginative, while Dimetrodon is an antic dowager. Will she last? Using unconventional materials and experimental, almost chaotic design principles, her builders were attendants at the birth of an organism that will continue to evolve over time.

Inventing a technology, and employing it for the first time, is always costly. There is the risk that the technology will fail (as the active solar space-heating scheme did at Dimetrodon) and the house will stand as a monument to that failure (Dimetrodon is redeemed because its successes far outweigh its failures). Better technology and better applications inevitably come as a result of the work of pioneers. Even twenty years ago, by distinguishing decisively between the verities (the need for good foundations and structure, the need for a storm-proof insulated envelope) and the cutting edge alternatives, Dimetrodon's builders found sound and timeless solutions.

Alternative energy innovation has not been exempt from a certain tendency to get it wrong the first time. An accusation quickly levelled at self-sufficient homes by the powers of convention is that their electrical systems are experimental, haphazard, temporary, and dangerous. Anyone who has rewired an old house may have encountered knob and tube wiring, the technique for electrifying (and burning down) houses during the first half of this century. A conceptually reasonable technique, it is difficult to service and performs poorly under normal stresses like rainstorms and cats in the attic. Newer approaches, based on observation of failures and pressure from the insurance industry, have evolved into the tech-

William Maclay and Dimetrodon.

niques we employ today. Electricians working with up-to-date materials and techniques are putting together safe, conservative, and ever-improving distribution systems; we know these will work well for decades, because the oldest applications of this new approach show no signs of decay after three decades. The electrician's task, and that of the electrical inspectors who are supposed to keep them honest, is made easier by their trade's maturity and years of accumulated experience. Changes in standard house wiring, equipment, and technology arrive more slowly now, and are integrated without strain.

Low-voltage wiring, for systems that encompass the whole electrical cycle from generation through distribution, confound conventional electricians and inspectors. Even if well-versed in technique and code requirements for grid-connected wiring, their theoretical grasp of whole systems and unfamiliar voltages may be shaky. Confronted by naked conductors, heavy-gauge wires, and a complex assemblage of unfamiliar devices and control circuitry, and not wanting to appear stupid, an inspector may make conservative grunts and mutterings and find a reason to withhold approval. In the early days of low-voltage residential systems, before a well-integrated system was feasible, many inspectors probably encountered rough-and-ready hook-ups, and concluded that alternative electrical practices are treacherous and shoddy. They have been slow to update their opinion. Fortunately, in areas where off-the-grid systems are becoming more common, inspectors are learning about alternative energy systems. Well-informed inspectors are right to hold us to a high standard of safety.

The wiring in some independent homes still looks rough; safety measures are not yet in place, and only the easygoing nature of low-voltage electricity keeps them from burning up. A strong case can be made that even the most unsafe alternative installations are safer under pressures than many high-tension utility distribution systems. Those who work with low voltage are often lulled by their discovery that DC is gentle compared to alternating current, but our trust in low voltage's forgiveness, and the habit of constantly revising and refining our system wiring, must not lead us to let down our guard.

As our reinvention of the independent home continues, a second wave of people approach the frontier. What pioneers made possible, the settlers will make permanent. Looking at the energy systems now in use, I see a second standard emerging; the pioneers' first-generation systems are experimental, and often mix and match primitive components in unusual ways, while second-generation systems can employ new, well-designed equipment made according to the same conservative standards as conventional house wiring. First-generation operators are innovators, and their methods are often eccentric, brilliant, and scary to those who do not understand them. As she or he works, the first-generation system operator tests, refines, and improves system performance—since this process will never end, why cover the system up? Insulation fatigue, conductor fraying and corrosion, and other problems will be found and corrected, because the system is continually being monitored, evaluated, repaired, and maintained. By contrast, the settlers want sophisticated systems that can be installed and, except for simple periodic maintenance, forgotten.

For some years, as low-voltage and residential power generation equipment advances rapidly, there will continue to be room for both modes. Since 1990 it has been possible to put together solid, safe systems using either. For innovators, new equipment offers opportunity and challenge, but for the new generation of independent homesteaders, the latest renewable energy technologies quite simply provide sensible ways to live comfortably without ever having to plug in.

Testing technology: Richard and Karen Perez's story

Karen and Richard Perez are among the original energy pioneers, and for many years have been evaluating shelter and energy systems in their high-tech, far-off-the-grid headquarters at Agate Flat in southern Oregon. On the way to publishing *Home Power* magazine they subject equipment to computerized testing with rigor and dedication; their accolade, "things that work" is the highest recognition an alternative energy device can earn.

Richard: We've been on this land for twenty-two years. I'm an air force brat, and I've lived all over, so this is my first ancestral home. This house started out to be my workshop, when I worked for a TV station up on Mount Ashland. Can I tell lightning stories!

Karen: I was a buckareena then, according to Sourdough George. I used to keep an eye on the rancher's cattle. They supplied the horse and feed and paid me. I was in heaven. I got to ride these hills.

Richard: We were fortunate, when we got here, to meet that old cowboy, Sourdough George. He was also a poet and philosopher, and knew a lot about the history of this area. The Shastas, Modocs, and Klamaths, our local Indian tribes, used Agate Flat as a springtime range. They dug apaw (a wild tuber in between a midget carrot and a potato, from which they made a starchy flour) . . .

Karen: . . . and ichnish, which is wild celery, very tasty when young.

When we came, nobody really wanted this land, and nobody wants it still. People leave us alone out here, except at hunting season. There are two hundred sixty acres in private hands out of, maybe, fifty miles square managed by the BLM.

We call them the Bureau of Land Mismanagement. The same guys that gave us the Food For Wildlife program, in which they released two-hundred-acre plots with a bulldozer. They put signs in to explain what they were doing . . .

Richard: . . . and the only thing growing there now is the sign.

We prefer the German attitude toward being foresters: you live in the forest you manage. Our guys have their office in downtown

Medford, the nearest big town, sixty miles away. There's a good story about a group of houses on Kennedy Creek in *Home Power* #20. Each of the first five houses carefully took enough water, without endangering the stream's natural ecology, to run a micro-hydro system that gave them all the power they needed (and some of them needed a lot), then passed the water to their downstream neighbor. The sixth house, which had the best hydro resource of all, belonged to the Forest Service, and they ran a 5-kilowatt diesel generator.

You can't tell it from the maps, but this is a great solar location. In fact, that's quite a ho-ho! for us all. In the Weather Bureau's solar data, we're between Redding and Medford, which is the third foggiest airport in the world. I think of the Nas Rudihn story about the student who finds the mullah scrabbling in the gravel by his front door, and asks, "What are you doing?" "Looking for my house keys," replies the wise man. The student joins the search, but after some time without finding them, asks, "Where did you lose them?" "In the kitchen," replies the teacher. "Then why look for them here?" the student wonders, expecting some enlightening insight. And he is rewarded; the mullah answers simply, "The light is better here."

The weather bureau looks at airports, I guess, because the light is better there, but local conditions can differ greatly in just a few miles.

Around here, we're often above the fog, and it's brilliant on the tops.

Karen: It's like a great white lake, with sunny mountaintop islands . . . Everything is a microclimate around here.

Richard: The Weather Bureau's data can be taken as a conservative guide, and you'll probably do better. Elevation is a wonderful thing for solar. It was incredible in Colombia at nine thousand feet . . .

Karen: The solar cooker worked even when there was cloud cover.

Richard: The sun oven was an important part of what we brought to Colombia, not so much because of the energy saved, but because it opened a door for the people we were working with: to understand this wonderful energy source falling out of the sky, and then they were eating from it.

Karen: We know that village women were quick to appreciate not having to walk three miles to gather firewood and hump it back to cook with.

Richard: If we make our own power from the sunlight falling on us, we become very sensitive about solar access. I expect that to become an issue like water rights, especially in urban settings.

We should also remember that power company workers are alive, breathing, concerned people with children. Our generation is late coming into control, and many of us are disenfranchised . . . All of us got into this—living and working the way we do—through the back door. The industry is populated by non-business types, and in fact everything goes well until the publicity and marketing types get in, like when the suits took over the computer industry a decade ago.

We found this place by getting lost, back on this muddy road with no way out, when we ran into Sourdough George. We chose to stay because we wanted out! We blame our present circumstances on the Grateful Dead: I got tired of changing batteries in the tape player. And the cats don't like kerosene.

Our first PV module was a big deal—hunger is the best pickle!—and we saved for it for a long time, but it produced hardly enough power to notice. It wasn't until we had eight that we felt like we had adequate supply. The constant fixes of new computer technology keep the system in a state of constant expansion.

Of course, we are always trying new technology. That noise in the background is the Survivor, a wind machine from Australia. Those Australians were amazing. They got here in the midst of the craziness of getting out an issue, and they dug the holes, poured cement . . . the president and vice president of the company installed it from the ground up. When the wind's cranking like this, we have more power than we can use.

We've got some concentrator modules on the test rack, but they require precise tracking, to within a quarter of a degree, to work properly. The idea is excellent: reduce the cell size dramatically, use satellite-grade single cells, and focus 150 suns on them. But to make them a consumer item, they've got to be cheaper.

Karen and Richard Perez check the weather outside the Agate Flat Proving Ground. The module testing line can be seen in the background.

When we moved here, twenty-two years ago, lifestyle was paramount, and what we did for a living secondary. Now, we don't have a life, we have a magazine! Having made a career up here, we couldn't leave if we wanted. Sometimes I think, let's move part of the operations downtown . . .

Karen: . . . and I say, No Way!

Richard: I get focussed on the work: I'll say, all we need to do is saw off my left leg and we'll be perfect. Karen's our conscience; she calls me to task, and says, wait a minute, you might need that leg.

Of course we're off-the-grid, so everything here is a laboratory, being worked on constantly. Our mission is implicit in how we live and in everything we do. Along the way, in our own experience and the stories we hear, we've developed some key ideas. Here are some samples:

Buy your land with cash. Don't make affording the land conditional on staying employed, or on the continuance of any economic circumstance.

Find old timers and ask them to tell you about the weather, special building cautions and opportunities. Sourdough George gave us insight into this land it would have taken years to find for ourselves.

Develop a second renewable resource besides solar. For us, wind is best, because the investment in civil works for hydro would not be paid back very well in the short hydro season. Wind matches solar perfectly, here: when's it's cloudy, it's windy.

You can always improve your storage. Find ways to store solar in summer. Be sure you understand all the inputs and outputs when buying an appliance. Lots of propane stoves, for example, have glow plugs that use 600 watts all the time the oven is on . . .

Karen: In Oregon, it's illegal to sell a stove with a pilot light because it uses too much energy.

Richard: And they may be right. But the glow plug is a real power pig.

Karen: You want piezo-electric ignition all the way around.

Richard: Don't tie your water system or your communications system to the main battery bank: give them their own fail-safe systems. Our radiophone has its own battery set, and our water has the pump connected directly to its dedicated PVs, so it pumps to the tank up the hill when the sun shines.

Ask for the help of good people. Karen is aces on getting necessary information over the phone.

[*There's a moment of anxiety as the sun fails, and the systems are checked to confirm there's enough power to make it through the night. Brutus is the name of the big, power-hungry sine wave inverter that runs Agate Flat's sophisticated computer equipment.*]

Richard: Brutus is still running. Can we tolerate a momentary power outage?

[*All agree, and Richard plunges us into the dark-of-the-moon blackness of Agate Flat without electricity while he turns Brutus off and shifts to a smaller inverter. The lights come back on and Richard returns from the control room.*]

It's gratifying, to use our own power, and to work with others that are concerned about energy. Larry, the Sun Frost fanatic, placed his refrigerator's coils in the box to reduce the condensation-desiccation cycle, so vegetables stay fresher. The Sun Frost door seals are a symphony of manufacturing, a continuous piano hinge that will last and resist the normal insults; the hot refrigerant is run along the door seals to keep them frost-free, where a conventional fridge would put heat tape.

Amazing thinking! But it is more amazing that so little thinking like that has gone on since we started burning all the dead dinosaurs we could pump up. The time has come for great changes. Even Southern California Edison's attitude is, no new construction should be without two kilowatts of modules and a synchronous inverter. As soon as we have adequate electric capacity, we can generate hydrogen, and move it around the same way we move natural gas. None of this is specialized technology. If the hippies in the woods can do it this well, then anybody can do it. We're in for a techno-primitive renaissance.

Beyond Experiment: Low-maintenance systems

Karen, Richard, and many other energy pioneers continue to devote their lives to improving energy technologies, but alternative energy specialists report an increasing demand for new independent home systems that require less from their operators. No longer experimental, such systems can be built with proven technology which is ready now for residential use. Those who live in experimental houses, and who are frequently asked to demonstrate their vanguard systems to curious newcomers, confirm the trend: the new wave of independent homesteaders are expressing a clear preference for durable, simply operated, self-regulating systems that incorporate the pioneer experience without requiring constant attention and refinement.

Native Americans, many of whom have never lived on-the-grid because their homes are too remote, and others of whom have moved to cities because their own land has no electricity, could benefit greatly from remote energy. But electricity has been employed as both carrot and stick in coaxing them to assimilate into anglo-civilization. Ethnic Polynesian Hawaiians find themselves in a classic good news/bad news situation. Good news: the state has set aside substantial plots of land on each island

Navajo home (photo courtesy Sandia National Labs).

for their homes. Bad news: health and safety laws decree that no building will be allowed on these Hawaiian Home Lands until they are fitted out with subdivision infrastructure in the form of roads, water, sewage disposal, and fire suppression. What will it take to convince the health and safety watchdogs that a home can harvest all the water and power it needs, manage its own wastes, and rest lightly on the generous Hawaiian land?

The second generation: Lafayette Young's story

On the island of Maui, a confluence of potential new home builders, a water shortage, and the energy crisis has created a vortex of second generation independent homebuilding. Laf Young is an academic builder and perfectionist who has applied his expertise to alternative energy systems. He lives with his wife Beverly in an owner-built, off-the-grid home on the ocean side of the twisty Hana Highway near Haiku. Both Laf and his son Cinco (Lafayette Young the Fifth) are finishing homes just down the road. Laf's new shelter is intended to be a vacation home for eco-tourists. While building for himself and immigrants from the mainland, he is always aware that his robust, self-managing electrical systems and self-sufficient catchment and residential water systems prove that a Hawaiian home is truest to its land when it lives without most of the burden of infrastructure.

Laf: After teaching vocational school in American Samoa, I came to Maui to administer the Community College, but after seven years I put myself back into vocational tech—homeowner's courses—so I've been exposed to all the building trades over my adult life. After awhile, you learn the right way to approach projects. I'm bringing in our new vacation rental at a ridiculously low figure, considering how well-built it is: under forty dollars a square foot, finished. I figure it will be a great vacation place for the guys who work with alternative energy all day, then go home to their all-electric homes. They need to live off-the-grid, catch their own water, to really understand the dream.

We raised our two kids here, and when they got home from school, and finished homework and the usual rural chores, they had to calculate, based on wind speed and battery charge, whether we were in a net positive power situation before they could ask to watch TV. A guy came over from SERI (the government's solar research institute) for his honeymoon, and looked me up at the college. I couldn't come out to the house with him, so I told him the kids would be home from school and could give him a tour. Cinco gave him the rap, and we didn't hear anything more about it until two weeks later, I got a letter. Apparently Cinco blew the guy's mind, because the letter said he knew more about energy and water systems than any adult at SERI.

Cinco Young explains the way he has metered his new house to his father, Laf Young.

Most of our power comes from an old Jacobs 32-volt wind generator. Do you know why they chose 32 volts? Delco was the market maker then, and they were looking to electrify the dairy farms in Wisconsin. You get good transmission above 30 volts, but the farmer standing ankle-deep in cowshit doesn't want too much voltage when he gets a shock. Turns out he can let go of 32 volts.

Deadly reliable machines, the Jacobs, turning 225 RPM [rotations per minute]. The low rotation speed keeps it from destroying itself in the bigger winds. I've seen a Windseeker go into run-away mode here, and spin so fast it melted the high-tech teflon tape on the leading blade edges. The Jacobs is easy for me to work on, but I rebuild motorcycles, too. I wish there was a good, modern, wind generator I could recommend.

We have a perfect wind resource, more power than we can use. Photovoltaic complements wind perfectly here, because the wind always blows when the sun doesn't shine.

In the 1970s when I had to go overseas, I'd call home and ask how's it going, and Cinco would say, "Shit, Dad, it's pitch dark in here!" Because the system was jerry-rigged, and we didn't have adequate controls. We didn't get balance-of-system equipment until the '80s. The best part of the Hawaii experience has been finding responsible designers, and keeping after them until we get the equipment we need. We've driven the industry to provide controls that just weren't available. I go away for weeks now, and the system just hums along.

It makes us feel smug, when Maui's grid goes down, and we're still living normally. For three years in a row, Maui Electric has gone down during the Superbowl; of course, the towers up the mountain lose power, and the TV signal stops, but friends call up to see if they should come over with their coolers.

In my original installations, now thirteen years old, the battery management isn't as good as it would be now, and the wiring is more California-style, wires looping cans together. Nowadays, I'm more concerned about managing the ferrous/nonferrous metals in advance, before they corrode and fail. Increasing the wire size between array and controls picks up unexpected efficiency: there's no point in heating up wires. In the past, stuff was so expensive that we installed the minimum, but now we should sell people the system they need.

I like to create bulletproof residential electrical systems that are transparent to the grid, so occupants and guests don't have to know. My philosophy is that I want everybody's experience with alternative energy to be a good one. The system should be self-maintaining; if you draw the batteries below a certain point, the generator should cut in automatically. There are no black eyes in this trade, but the first house that burns down because of crummy DC wiring will be a tragedy. I don't mean it to be one of my houses.

Architecture in Hawaii is hideous. No one is considering the energy future. Even the energy present is bleak, because we import all our electricity as petroleum. I like to get customers cold, before they've been spoiled by the architects. We tell them about how Hawaiian houses can cool themselves by letting the trade winds in. Fans don't like inverters, and speed controls and inverters are a nightmare, so it's better if we let the air move itself. Usually, the guy says, give me a 24-volt system; I'm ready to go, but when I ask the wife if she wants a washing machine, they both look horrified. Of course they do! By then, I'm thinking about a 24- or 48-volt system. I get people to come with a list of their present appliances, and we work backward from that. I get some of the best questions from the women, because they really grind on the concept until it gets clear for them. Beverly's helpful in the summer, when she isn't teaching first grade, because she can show folks around, the Maytag and all.

When I've worked with a malahini (a newcomer to the Islands) and put in a new system, I figure on two call-backs after they're launched. Either there will be a few operational questions, or they'll crash their batteries through inattention. The hardest part is keeping the system balanced, and that's because it's hard to measure the battery's state of charge. Voltage is no good, and neither is the hydrometer, because unless you're doing something to stir the electrolyte you've got stratification. My experience is that if a system is well-thought-out, it seldom fails, and so people don't lose refrigerators full of food. Then I'll see them on the street or in the supermarket, and they'll come and give me a big hug and say, "Thank you, you've changed our life!"

It's important to understand the whole scope of the system if you want to live inside it. If you're already used to an electrical lifestyle, you need to retune, and ask questions you've never asked before: Where does this come from, what does it cost, and do I need it? When you understand that you are catching your own electricity and water, you start to practice water and energy conservation naturally. Instead of being upset when you get a rainy day, you say, Yes!—because you're catching water and filling your tank .

The most serious mistake to make in the tropics, especially in the low altitudes, is to build on a slab right at grade, so there is no air underneath and all sorts of insect and critter problems ensue. Centipedes are a specialty (one got in my bed just the other night), and higher up, there are scorpions. In our climate, we want circulation over, under, and through. The climate is so hard on wood that the roof should carry well over the decks, which should be wide enough to use.

The tropics are a difficult environment, but you take the same care in building you would elsewhere. We foam the holes drilled for electrical wires to control infiltration, but it's insects, not cold, we're controlling. For the same reason, everything below the roof is treated. If there's an open or cathedral ceiling, it makes sense to insulate to keep it cooler in the daytime. Cross-ventilation, and orientation to the trade winds, is critical. I used stainless steel screws on my rental's new roof because aluminum screws back out, and I'd rather spend a thousand on screws

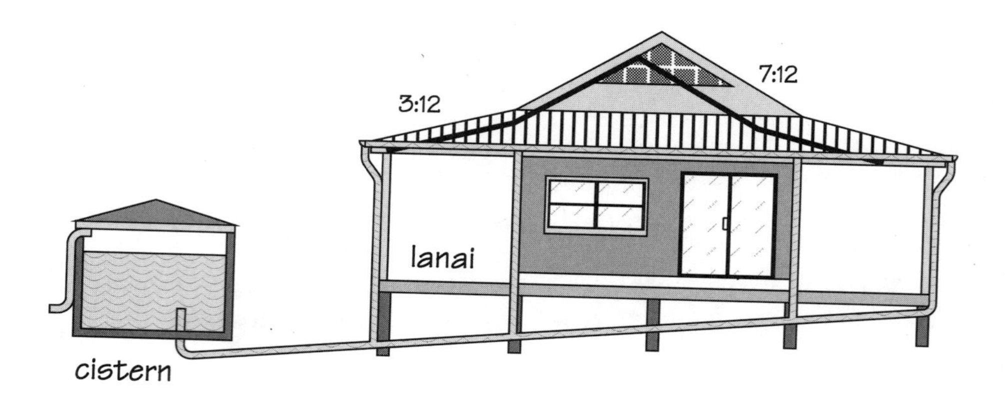

Figure 36. C. W. Dickey's graceful proportions work well in the tropics. All rainfall on the roof of this house is piped to the cistern.

now than several thousand on a new roof after a storm. I like C. W. Dickey's split-pitch roof for the Islands, a seven-in-twelve pitch over the peak, then three-in-twelve over the decking. I've found that, with a little blocking, I can do it without breaking the metal roofing where the pitch changes. We're at 21 degrees of latitude , and so a four-in-twelve pitch would be better for solar panels, either water or electric, because you need to optimize for winter here. At the new homesite, we get stellar afternoons, so I oriented the house about 10 degrees west of south.

On Maui, if you need remote power, you need remote water, too. On the windward side, there's plenty of rain, so a good catchment system will do. I gutter the roofs carefully, and bring the downspouts to a manifold which comes up inside a good-sized ferroconcrete tank. Residential water should be reliable, so I don't like the lightweight, intermittent-duty pumps. A flo-jet out of the box will fail here in a month or month and a half because the contacts corrode.

Of course you use solar hot water. In Hawaii, the natives (whose land all this was, originally) have gotten land grants, called the Hawaiian Homestead Lands. There are two big plots here on Maui, and every native is supposed to get a parcel. But the bureaucrats won't let them build because there's no infrastructure. They've got to have water meters (and there's a moratorium on those), fire suppression, paved streets, electricity. I'm trying to build the case that we work around those problems for rich haoles [non-natives] all the time, why not for the original residents?

I've also been lobbying the banks to make loans for alternative energy. I've finally gotten First Hawaiian Credit to agree to finance a responsibly installed alternative energy system. It's very hard, maybe impossible, for most lenders to write loans in a rural environment when they encounter alternative energy, because all banks resell their notes in a banker's market that requires conventional housing, you know, streets with curbs, powerlines. If they can't resell, they have to hold the notes themselves. The Community Reinvestment Act (CRA) requires them to do just that, to keep a percentage of local notes. My approach worked, because I explained to them why alternative energy projects were good candidates for their CRA portfolio.

I spent thousands of hours and dollars lobbying the whole state to allow 6-liter [1.6 gallon] flush toilets instead of the wasteful three-and-a-half gallon models, and they're finally legal. At one point in my campaign, the plumbing honcho told me he could get them legalized if they came in some color other than white. I asked what colors he had in mind, and he said, "Green . . . ?"

The PURPA program here is a joke. They'll buy your excess power for five cents a kilowatt then sell it back at thirteen cents retail. Then they charge you a fifty-dollar monthly metering fee because they say it takes a second meter reader with special training. So it doesn't happen much. I understand the utility's point: on an island, where it is the sum total of its own grid, if it can't dispatch a resource, it's a frill.

Committee work can eat you alive. Friends of mine spend hours on IRP panels [Power company Integrated Resource Planning advisory groups] and it's all voluntary. I'm on sabbatical from social involvement, and Beverly has taken over. She's working to reorganize the Education Department.

Laf Young on Gas Refrigerators

Gas fridges work by absorption. They don't make cold, they remove heat by exposing the chamber to a mixture of liquids and gases that draw off heat and then radiate it through fins on the back of the box. The evaporator in the ice cube tray must be exactly level, because the droplets of refrigerant run down by gravity. Some units had built-in levels: that was a good idea. The newer models are sometimes so poorly made that levelling the case isn't good enough, because the tray is off level, so be sure to level the evaporator inside the box by levelling the freezer shelf.

If a gas refrigerator stops working, see if it will run on electricity. If it does, you know the burner is dirty or the chimney is clogged. If it doesn't, and you verified that you really have given the unit the right kind of electricity, check the temperature of the refrigerant reservoir at the bottom. If it's hot, turn the unit off, quick. If it gets too hot, the buffering chemical in the hydrogen-water-ammonia mix cooks into a solid that plugs the little tubes in the evaporator permanently. The old idea—that you could turn the whole box upside down—doesn't work with the new units. If you find green goo around any fittings, or you've smelled ammonia, then you've had a leak and the unit needs recharging.

You have to service a gas fridge once a year. Do it yourself with a little homemade manometer with the right kind of fittings on the ends. You're looking for eleven inches of pressure. Clean out the fur balls and the industrial ruby orifice, blow out the burner, and take a shotgun brush to the heat exchanger. Otherwise, call the service guy. It's cheaper to maintain it than to replace it.

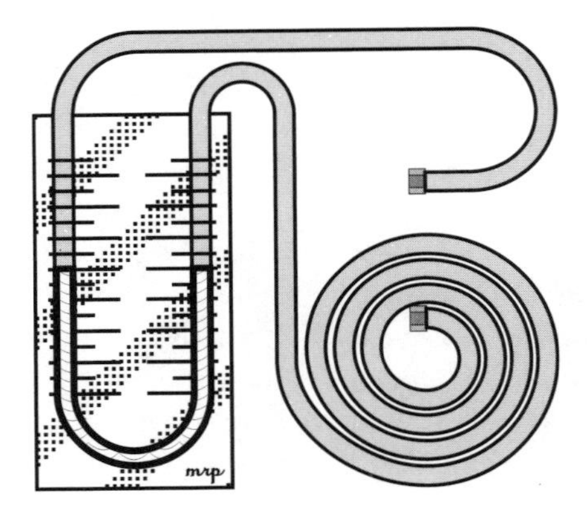

Figure 37. Laf Young's homemade manometer consists of clear plastic tubing fitted out with the right brass connectors for your gas refrigerator's plumbing.

I'm worried that technology has divorced us from knowing how things work. Like semiconductors: What do they do? And is there anything we can do about it when they don't? My son-in-law is trained to troubleshoot equipment seven levels deep, but when the problem is found they still swap out whole modules, not the failed component.

We trust that technology will take care of us, but actually that means we need to have (and trust) an elite, who do understand, to take care of us.

Integrated Response

The secret of an independent home's success is the integration of separate systems into a responsive machine. Where a conventional state-of-the-art unit (like a heat pump, which incorporates electrical, plumbing, and mechanical subsystems) offers an opportunity for the electrician, plumber, and mechanical engineer to point their fingers at each other, the independent homesteader expects a more holistic approach. Conventional systems are meant to be stuffed inside walls and forgotten, but the independent homeowner expects to be involved in the day-by-day and hour-by-hour management of the home's systems. The more mechanistic image of the home is softened by the intimate inclusion of humans in the management loop. As Michael Reynolds has said, the independent home is a sailboat, not a motor launch. The power-gathering, heating and cooling, lighting and control systems work best when they are steadily harmonized and adapted. Once these systems are in place, the designer must pass along the knowledge essential to their management. It is hard to know if Laf Young is prouder of his house or of his family's ability to operate it happily in his absence. In my view, only part of the credit for the success of renewable energy goes to improved systems; the pioneers have been heroic in their dedication, and it takes years of conscientious caretaking for a family to learn what it has taken generations to unlearn, by neglect.

Three conditions contribute to "bulletproof" residential systems: the system must be robust and clearly informative, so the home's human collaborators can understand what they must do to optimize energy production and conservation; adjustments and periodic servicing requirements must be manageable and self-explanatory, so the operators will be eager to take an active hand in operations; and the operators must know that competent help is available, so they are encouraged to be curious and ask questions. Good design, lucid metering, and decent transfer of technology from installer to operators will satisfy the first two requirements, but it may take another decade before responsible local support is available everywhere it is needed. In some locales, pioneers have taken their expertise into the marketplace and adapted their techniques to the needs of the newer settlers, resulting in a happy and thriving energy self-sufficient communities. In other regions of the nation, pioneers are proving that alternative energy systems work well, but a community of settlers has not yet taken root around them.

Knowing what I do about the trials and travails of independent homesteading, the homes that appealed to me most, that felt like home, were built for the long term. In these places I could tell without asking that important, easily overlooked details had been well-handled: details like stormproofing (minimizing infiltration, confining roof leaks to low-tech areas, arranging for reasonably mud-free access) as well as health, safety, and disaster planning—including cross-training and documentation, so the principal operator can go away overnight without the house crashing.

A pioneering shelter may be little more than a lean-to for protecting dry goods from the storm. Jim Loomis, who lives in Maui's jungle and considers himself a water ape, asks only this of his shelters: that they keep most things mostly dry most of the time. For explorers breaking trail into new territory, it may be this kind of deprivation that keeps the pioneering edge honed, and I admire them for their fortitude. I consider myself a settler, and set for myself and my family a more convenient standard.

As I visited pioneers on the technological frontier, I noticed again and again that the work was so consuming that it inevitably spilled over into the life, to the point that work would engulf life without an almost epic effort to find a balance with more traditional endeavors. (Richard Perez's comment, "We don't have a life, we have a magazine," and Dick Britt's story of Sardo winning his spurs come quickly to mind.) As these energy frontiers are tamed by the pioneers, however, the renewable energy life becomes possible for settlers like myself.

CHAPTER 10

HOME ENTROPY: *Improvements and Repairs*

ONCE COMFORTABLY ESTABLISHED ON THE LAND, WE INDEPENDENT HOME-steaders intend to stay. Changes there will be: we may add rooms, batteries, or photovoltaic modules, and those who started far from the grid may even connect themselves to it if the power lines advance close enough. But we are not leaving. I cannot imagine living anywhere but Caspar, or in any other house. This life is more than place, it is process; the land, the home, and I grow together.

It might occur to you that people who live in excruciatingly beautiful houses are trapped. They can never leave, because there is no place to go as nice as where they live. So if you have an investment in being rootless, be very careful that you do not live someplace too nice.

Those who are content to be trapped by a happy home place develop a good sense of what it takes to maintain a home, and a concern for repairing its failures and improving the way it works. As time passes, most of us formulate strong conclusions about how a house should be put together so it stays together.

Living at the end of the road, and relying on unusual equipment to generate our energy, who can we call when the music stops and the lights go dim? As more and more people locate themselves in the backwoods, three things will happen. The equipment will get more reliable. The service infrastructure, trained workers with the right equipment to maintain and repair whole energy systems, will improve. And, because the repairers will be reluctant to come to the end of the road for a triviality or on a Sunday, the people that live there will get better at diagnosing and maintaining their own systems.

Spreading the news: Richard Gottlieb and Carol Levin's story

We are often called upon, in this age of rapidly changing technology, to cope with equipment and concepts for which we have never been formally prepared. It is barely possible in this complicated world to stop learning. In earlier centuries, when the pace of innovation was much slower, it was normal to stand by tried and true ways, but most of us now learn perpetually. Some of us can find what we need in technical books and users manuals, while others need more structured situations, a classroom and a teacher.

Richard Gottlieb and Carol Levin live and teach solar energy techniques in a rural valley near Brattleboro, Vermont. Their photovoltaically powered house and workshop gets its back-up power from the grid.

Carol: I got into all this solar energy business because I met Rich when I was running a coffeehouse in Brattleboro and we got married.

Richard: What galvanized us into doing solar was our first experience with a window project, where we grossed half as much in a week as we had in the previous half year's farming. We started doing solar hot water, back during the tax credit era, but there was lots of competition, and then the credits evaporated.

Carol: Solar is like magic, you put up panels, and electricity comes out. We travelled some, visited with friends in Tempe, Arizona, who said, "You can't do solar in Vermont: that's in the blue zone."

Richard: You know, referring to the map that shows solar potential, it's red in Arizona and yellow part way up the Atlantic Seaboard, and blue says there isn't enough winter sun to make it work. We thought, "solar is a neat idea anyway, and there's nobody doing it in New England," so we started doing hybrid systems.

Carol: We just got into our niche, put in our own PV system, started gathering data, and discovered you could get plenty of power here. We were petrified when the tax credits went and hot water went down the tube, but then we found that where people wanted solar hot water for the tax break, they wanted PV for its value. And they were wonderful. Where else would you find people who invite you over for a champagne party when you finish putting in power to run their home?

Richard: We make people happy. Some of the people we work with are making a philosophical statement. Maybe 20 percent of our customers want to face their grandchildren when asked what they did to keep from ruining the planet. They come to us and ask, "What can we do?" So we tell them. More and more of them are really having fun by making a statement.

We are most pleased with the work we've done with Ben & Jerry's bus and truck, which are seen by thousands of people everyday. They've got solar panels on their roofs, which crank up when they park. They keep their ice cream in Sun Frost freezers powered by energy from the sun.

Every year we provide solar amplification for Pete Seeger's Clearwater Hudson River Revival Festival. We set 300 watts of modules on top of our van, take along the 200-watt tracker array, and plug in a 500-watt sine wave inverter to the batteries that power the old-timey stage.

We run one-day, hands-on classes for eight people every four or five weeks from spring through autumn right here in our workshop. Theory in the morning, then we assemble a four-panel system in the afternoon.

Carol: About half our students are homeowners, a quarter professionals—architects, electricians, contractors—and the rest are students and retired folks who are interested in seeing something new. Most of them have never done any electricity before, so we start out slow.

Richard Gottlieb and Carol Levin in front of their workshop.

Richard: Some have had more experience than others, who may hang back, but on the panels there are eight connections, so everyone gets a chance for hands-on experience. By the end of the day they see a whole system, panels, transmission, batteries, load center, and equipment doing something useful. And they get confidence that they can do it.

Mostly, we do off-the-grid projects around southern Vermont. There are 500,000 people in Vermont, and six dealer-installers, and as many PV systems per capita as in Arizona. Here in Vermont there are lots of regional power companies, and the electrical source may be 60 percent hydro, 30 percent nuclear, and 10 percent other, much of which is bought from neighboring power companies through the New England Power Pool.

Carol: Most people come to us after finding out how much it costs to bring in a power line. They just want lights, a refrigerator, a pump, entertainment, a computer, and they find out it will cost five to eight dollars per foot. For that much money, it's easy to break even on an independent system if the extension is half a mile or more.

Richard: People come to us with the line extension cost as their price: "What can I get for fifteen thousand dollars?" With help, most people get the installation right the first time, although we often see the systems grow as soon as people see how easy they are. There's a constant tuning process, as new appliances get added, or they add panels, or the trees grow and the tracker needs to be moved. We really try to anticipate where they will want to be three, five, eight years after installing their system, and put them into the right equipment to begin with.

When we put in a 500-watt system, installed, it costs about thirty dollars a watt now. If they assemble and install it themselves, it comes in at about half that, fifteen dollars a watt.

Carol: Or we'll preassemble it here for something in between, and they do their own installation.

Richard: When we do the whole system, we make sure it conforms with the spirit of the National Electrical Code™, and we walk the new owners through the maintenance, and give them a manual: a nice package. We've done a few nice systems for people who can get grid power easily, but prefer to be their own power companies.

Carol: We also get people who have been living away from the power lines, and are tired of kerosene lights or they've got kids coming up, and they want proper light for homework. There's one guy, a welder from Putney, who empties the change from his pockets every night when he gets home, and every year or so he's got enough to add another module. And then there are some long-time customers who have been running their systems for seven or eight years, and it's time to change the batteries.

Richard: Nobody wants to water their batteries properly. When the Ben & Jerry's Solar Roller came back after two years, all the batteries were practically dry, but we'd sized them conservatively and they were still working. I put two gallons of water into them. But with proper training and attention, batteries are not much of a problem.

Some people are starting to worry about electromagnetic radiation (EMR). I don't know why, but we started working with DC lights from the beginning, and have gradually switched most of the lighting in our house over. I just don't like AC lights.

One thing that comes with experience: we find ourselves doing prettier systems, with cleaner work. The equipment is better, easier to work with, and the National Electrical Code keeps changing. We closely follow the work John Wiles, the safety guru, is doing with proper fusing and grounding, and incorporate that in our systems.

Carol: In Vermont, the state cares about septic systems, but there's no building inspection outside the big towns, and no agency cares if you build things right. Our own system is always under development. We'd like it to be state-of-the-art, but customers' needs come first, and so something's always under construction here.

Richard: But the days of alligator clips and jumper cables are over. Independent power systems are in.

System Failure: Thinking like an island

When the power fails in an independent home, for whom does the bell ring? Right: for us! We are the power company. It is a well-documented fact that grid power fails more often than independent power; organizations like the Federal Aeronautics Administration and the power companies themselves use reliable independent systems for back-up, for equipment that cannot be allowed to fail. Independent home systems seldom fail, but they may dwindle. If we have slighted periodic maintenance, and not been attentive to our meters, this may come as a shock the first time or two, but then we usually learn to pay attention, and head off trouble before it happens.

Over the life of a house, mechanical features, anything having to do with wires, pipes, or motors, will someday require maintenance. The practical rule, a corollary of Murphy's Law, is that the probability of failure is directly proportional to the difficulties involved in correcting the problem. Problems encountered while repairing other problems can change our lives in spectacular ways if major reconstruction becomes necessary. I recommend placing removable panels over walls that contain mechanical, plumbing, and electrical connections, so that systems may be revealed and repaired and the panel replaced as neatly as new. Screws are much less expensive than nails when they must be removed.

There is no substitute for an intimate, up-to-date knowledge of a home's systems. Experienced homesteaders will catch and correct problems before they turn into outages. To avoid the "it only fails when I'm gone" syndrome, the daily inspection responsibilities should be shared by everyone in the family tall enough to read the meter, evaluate the water flow, or note any other changes that may reflect on system health.

By refining our sense of what should happen to a system, and comparing it with what is actually happening, troubles can be anticipated. Experienced users have a finely tuned sixth sense of how much sun fell on the panels, and how much energy should have been accumulated in storage. If the actual storage state is less (batteries are seldom found to be fuller than expected), then a short, seasonally adjusted litany of causes is checked. For example, in the fall, as the sun's apogee falls lower in the sky, new growth in trees may shade the photovoltaic array where a year ago the array stood in full sun. A failing module, loose connections, a faulty charge controller, a power-hungry task performed by another family member, a small load left connected: all these can change the state of charge. A single variance is not a cause for concern, but if the system continues to perform below expectations, the problem must be found. This implies vigilant attention on an almost-daily basis. Since this reckoning is complicated by season, climatic variations over many years, and changing supply and demand, new operators should not be despair; the intuitive sixth sense of the system takes years to develop.

The amp-hour meter has revolutionized the way we track our power systems by showing us the balance between incoming and outgoing power, letting us know exactly what energy we have made and spent. As power is generated, the meter counts up from zero at the battery's discharged state; as we use it, the count goes down. In a home power system, you cannot run on credit. Once we learn how much electricity a task requires, we can easily see if there is enough energy in the bank. Preventive system management minimizes surprises.

Failures in new systems, where the operators have no experience to rely on, are surprising and confusing. One reassuring difference between electricity and water systems is that electrical systems, when they fail, do not fill rooms two inches deep in escaped electrons. Installed safely, power fails safely: it stops, and the lights go out. The first step when the lights go out is to take a few deep breaths and think about causes and consequences. Grid-connected folks cast their imaginations farther afield. (Did another drunk hit a power pole? Did a tree fall on the transmission line? Has a terrorist attack begun?) The independent homesteader need only think through the local system and the way the power failed. Did the lights slowly dim, or did they go out as if switched? Which systems are down? Were there any sounds or smells? Where is the flashlight?

Water systems have their typical failure modes, too. In a gravity system, if the source has failed, pressure decreases slightly with usage, then there is no more water. In pumped systems, pressure decreases from normal to a dribble as water is used. In either case, hours may pass between a failure and its recognition. Salamaders periodically stop Frank Dolan's system, and earwigs in the pressure switch account for 80 percent of my water system failures. By using each failure as an instruction on how to make a system more robust, we may expect failures to become rarer and more interesting. To combat the earwig problem, we have installed a small lightbulb inside the pressure switch housing, which seems to keep the photophobic insects away.

When installing an electrical system, it is a mistake to economize on metering, switching, and lighting in areas that must be repaired. This applies to plumbed systems, too: careful use of valves and unions let us repair rather than rip out and rebuild. Time and money spent in making systems easier to understand and operate pays itself back quickly, often the first time the system fails. Good systems are designed to work well, to be maintained and repaired easily, and to recover quickly from failure.

Well-organized systems are switched, metered, labelled, and documented so that it will be easy to localize a failure. Once the failed device is found, we should know, or be able to find in the system manual, what procedures to follow. Before the system really fails, supervised "lifeboat drills" will help us know what to do. It is easy enough to separate essential circuits—minimal lighting, water pumping, telephones—and make sure everyone in a household who must operate the system knows how to shut everything else off.

Critical components—computers, water systems, telecommunications—should have their own back-up power or be on separate circuits so that we have a few minutes to collect ourselves and make an orderly retreat after the lights fail. A delicate balance must be found between too many redundant systems (each of which require separate, time-consuming maintenance) and too few self-sufficient systems (that can function when other systems fail). My priority is to let water pumping go, but to maintain computer power at any cost, although others who share my land consider my priorities to be a bit askew.

Independent homes should be well-endowed with functional flashlights, easily locatable, near places where people spend time or where darkness would be disconcerting. Twelve-volt lights located at places in the system where troubleshooting may be required—the metering station, the distribution center, the battery enclosure—make it much easier to find and correct problems.

Failure Modes: Darkness or dimming?

System failures come in two types: device failure (rare) and empty batteries. The first step in successful troubleshooting involves reviewing the events leading up to the failure and the circumstances at the time. Two things help track down the problem: an understanding of how energy flows through the overall system, and the ability, through well-conceived metering, switching, and overcurrent protection, to isolate flow-paths and test them. It is simply a matter of thinking of particles, and following their flow, until the stoppage reveals itself.

If we imagine the system as a stream of energy flowing from its source through storage to the loads, we readily see that a failed device will normally interrupt the flow to all downstream constituents, which go dark. If something was just plugged in or turned on, this is the likeliest culprit: either it is defective or it drew too much current and a protective device, a fuse or circuit breaker, shut it down. In grid-transparent installations, where house current is delivered through conventional plugs and in-the-wall wiring, failure may be so infrequent that we forget there is a home system present, and commit a momentary lapse in prudent load management. A sure way to cause this kind of failure is to plug in the iron while the washer is agitating. Check the breakers and fuses, and vow to manage better in the future.

Sometimes system dimming takes place over such a long period that we do not notice (or think advancing age is causing our vision to dim) until a voltage-sensitive device notifies us by invoking its low-voltage protection and unceremoniously shutting itself off. This is a strong message. In the future, we should monitor the system's meters to be sure we are the first to know about low system power.

Whatever the failure mode, the place to start troubleshooting is the affected system's center and likeliest trouble source: the battery bank for electricity, the pump for water. We should be able to check battery voltage from the metering station. If it is low (in a 12-volt system, that would be below 12 volts), there are two probable causes: a big load is drawing the voltage down, and should be disconnected as soon as practicable, or the batteries have been asked to supply more electricity than their sources have provided or they have been able to store. If a big load is causing a low-voltage brownout, removing it may solve the problem temporarily. Even when the problem is an overdrawn battery bank account, and the batteries have been drained down below safe operating levels, it will generally be possible to get through the night by minimizing loads until the batteries can be recharged. If the big load is important, it may be time to start the back-up generator.

If the batteries show plenty of charge, then the problem is downstream in the power conditioning or distribution systems. Inverters are the most complicated technology in the system, and so are next in line, after the batteries, as candidates for failure. If alternating house current is out, but 12-volt loads still run, check the inverter. Most inverters have built-in overcurrent protection, which may have tripped; remove all AC loads and try resetting the inverter's circuit breaker, then test the AC output with a small, expendable AC load like a trouble light. If all seems well, gradually reconnect your AC circuits.

A trouble log, perhaps kept as part of the system operations manual, provides a useful record of problems, and will be helpful in improving system management. Utility companies are fanatical about their system logs, and for good reason: we learn from subtleties and patterns which may only be retrievable through an orderly practice of logging problems and solutions. Good records will help a troubleshooter localize problems, and will provide family members with a summary of previous experience.

Figure 38. System log.

System Log	
starting date:	1 Jan 1992

date & time		who	observation
1 Jan	3:00pm	mrp	watered batteries (used 11 oz); 4 days of rain, batteries low
3 Jan	12:53pm	mrp	12 minute grid power failure; batteries low 2.73" rain
6 Jan	6:30pm	rfe	2 sunny days; batteries full, fan running
8 Jan	11:15am	smp	big storm; general power outage max gust 83mph
	4:25pm	smp	no water pressure; power still out, TV & VCR OK on battery
9 Jan	7:45am	rfe	power restored; lost prime on pump remedied 1.19"
11Jan	9:45pm	rfe	no water pressure; removed earwig from pressure switch
12 Jan		mrp	Francis installed neon light in pressure switch housing 0.15" rain
13 Jan	6:00pm	mrp	sunny still day; max out temp 83.5oF, indoor 75.5 w/ no fire

Optimizing Performance

As soon as we find ourselves within new walls, either new-built or new to us, we begin to find innumerable ways to make the house work better. We seek to achieve a transformation that is alchemical, to turn the house into a home. By lavishing intention and love, by investing ourselves in the walls, fixtures, furnishing, and all the fine-drawn details of a shelter, we will work that alchemy.

Any plan is limited by the fact that we cannot always see to the end of a process from where we stand at the beginning. Attempting to visualize and predict seldom gets every factor right. By actively participating in construction, we find that our vision of the home shimmers between intention and reality the whole time it is being built; as a room springs from floor plan to enclosed space, we may discover unexpected views and delightful corners which alter our expectations about how we will live within the space. The resulting alterations can usually be accomplished better and less expensively during construction than after the house is occupied. Meanwhile, as we build, under pressures of time and money, we will defer certain other enhancements and leave certain other puzzles unsolved until occupancy, knowing that daily living with the home will make the solutions moot or obvious. Inevitably, some possibilities are missed along the way.

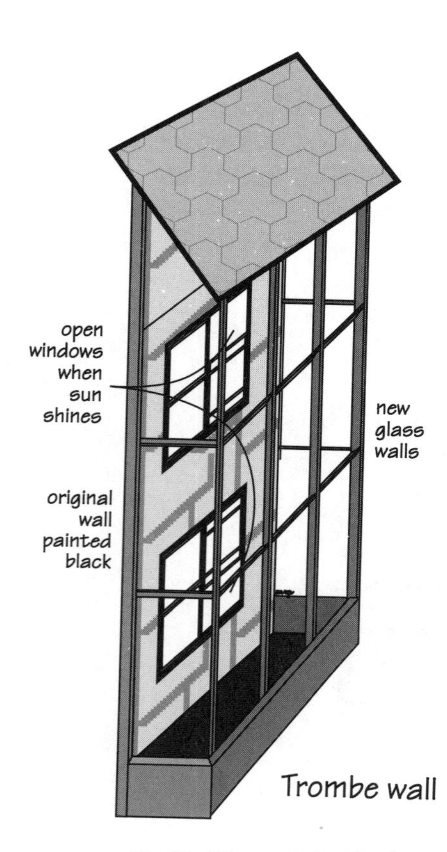

Fig 39. Warm air inside the glass trombe wall, heated by the sun, rises and enters the house through the upper window and is replaced by cooler air flowing in from the lower window. At night, the windows are closed.

Many of my own first efforts were waylaid by inexperience and by my willingness to plow ahead uninformed. Still, most of my early results were as good or better than conventional solutions. Many independent builders praise this experience of stepwise refinement, in which we are willing to try, enjoy limited success, regroup, and try again.

As patterns develop, as we find the places we prefer to sit, work, and move through, and as the furniture settles into its right locations, enhancements begin to recommend themselves. As Robert Sardinsky told us in chapter 4, lighting is functional and easily ignored, but lighting makes more difference in the habitability and friendliness of a place than any other single element. Lighting is also a major energy consumer, and therefore careful choice and placement of lights for tasks makes a dramatic contribution to efficiency. For daytime activities the least expensive light, including its heating and cooling implications as well as its illumination, is proper window placement.

The membranes of a house, the walls, roofs, floors, need not be thought inviolate—set in cement, as the saying goes. Opening a window through an existing wall may not be such a daunting task, especially if the window is small and fits between the wall's framing members. The resulting framed view or passive task lighting quickly repays our effort. Architects insist that a variety of kinds and shapes of windows "look funny." We should be concerned, I believe, with how windows look and work from the inside, not how they are perceived by neighbors and wandering architectural critics. The odd assortment of windows in my house look out wonderfully on lovely surrounds.

Like any other appliance, windows can and should be replaced when a better-performing appliance, like a high-tech, low-emissivity window, becomes available. The cost of replacement will pay for itself very shortly, and continue to give efficiency and satisfaction indefinitely thereafter.

Moving conditioned (heated or cooled) air is a simple and effective energy-management project. Well-developed convection patterns designed into the home from the planning stage are best, but retrofitted registers, ducts, and other passive measures can improve circulation when the original plan does not quite work. Small fans, possibly directly connected to solar modules or running on surplus energy, can help redistribute trapped pools of expensive comfort. Tropical slow-fans accomplish this and add an aura of intrigue to a home, but small openings between rooms and between floors can accomplish as much less showily.

It is very cost-effective to add well-conceived energy features to existing homes. Maria and Arnie Valdez have helped their neighbors build passive solar collectors onto existing structures in the cold high desert of southern Colorado's San Luis Valley. The original builders might complain that these retrofits are graceless, but the inhabitants are warmer, their energy bills are lower, and another declaration of self-reliance and renewability has been made. There is a nobility and commitment to such additions: herein dwell people trying to live well by adapting to new conditions. Partly in response to their example, future buildings will be planned to incorporate these features from the beginning.

As soon as we understand the meaning of the pipes, ducts, and wires in the walls, their origin and purpose, it becomes easy to adjust them to our needs. If lighting or electrical outlets are located inconveniently, a solution more suitable than extension cords or deprivation must be found. Permanently adding outlets and fixed lights can be a more complicated task than adding a window, because a section of wall must be removed (or worked behind) and then restored. The best solution, of course, is to anticipate the need and put in adequate fixtures before the wall is panelled. Next best is to employ a tolerant panelling scheme where changes may be required: plywood rather than gypsum wallboard, screwed to the studs rather than nailed. I have successfully installed remedial wiring in built-in cabinets and counters, and surface-mounted wiring hardware is an acceptable last resort. Adobe buildings and earthships are surprisingly tolerant: to add an outlet, one simply chips out plaster and adobe to about twice the depth of the wire and box, sets wire and box in place, then replasters and repaints.

Never Finished: Enlarging on the Theme

One way to make sure that a home adapts as well as possible to our needs is never to complete it, but to see the home as always growing, always undergoing substantial change. This institutionalized incompleteness was mentioned as benefit and a regret by many who live independently; for some, who had lived amidst construction debris for too long, the best argument in favor of a conventional home was the attraction of living in finished rooms. Neither way proves to be exactly right for everyone; what we must decide is which of these polar opposites we wish to steer toward.

Built solidly of wood and stone, copper and glass, houses nevertheless are living organisms, and tend to grow. Under pressure to design a home within unyielding time and financial limitations, we rarely allow for more space than we will be require. Most often, we fail to anticipate social and familial metamorphoses which, in a relatively short time, alter the way we relate to our home space. One trend is clear: I visited no home where space was being removed, and in most of the homes visited, an expansion of the enclosed space was being actively planned or was underway.

I do not want you to think that an independent home can never be finished. More than half of the homes I visited were substantially complete, and any additions or improvements were small and being undertaken with the quality of immediate life very much in mind. The idea of a home as an elastic, malleable, growing, and adapting membrane to live inside is dynamic and unique, but changes to enhance life in the future ought not put too much strain on living in the present.

Whenever an addition or improvement pierces the membrane or requires heavy construction, we must endure the hardships of living in a construction site. Savvy homesteaders have learned to minimize the dust and incompleteness (because life must go on) yet we are often lax in applying our knowledge. My experience proves—even if I persist in ignoring it—that a well-planned temporary containment for any building project repays the time spent on its conception and execution in terms of less dust on every horizontal surface in the house, less sawdust in the bed, and fewer indoor gales and deluges. Take the time to spread dropcloths, close off doors or zones with plastic sheet or tarp, and clean up after small but messy project segments like sawing and sanding while the dust is still a local phenomenon.

One way to manage our need for ever-expanding space is to build another building. Direct connection through a breezeway or attached hallway works best. David Palumbo's shop communicates with his big house through a storage hallway. The glass greenhouse connecting the power room and privy to Frank Dolan's main house also serves as a canopy over the front door. These connective structures provide additional storage and protect their residents from inclement weather. Starting over with a new building recommends itself for a number of reasons, not the least of which is that we get to do it right based on previous experience; no longer exactly naive, we may look forward to making a whole new set of mistakes.

Customize or Homogenize

Assembling quick shelter is the imperative of most modern construction. In homebuilding, speed is too often the antithesis of quality; the closest fast residential construction can come is a sort of mass-produced gingerbread totally lacking in the spirit and pride of the traditional. We are not fooled, but remain aware that shelter that goes up fast can come down fast. Because of the tempo of hurried development, a homogeneous building style pervades much of America. All these buildings share a common ancestor: the warehouse. Independent homebuilders are not afraid to appropriate the best tools and techniques of mass-production building (nail guns, rigid foam insulation, high-tech glass) but they employ them with the craftsman's spirit.

Again and again, two themes were evident to me as I visited independent homes and their proud owners. One theme, harkening back to homebuilding in earlier centuries and to the spirit which lives on in fine furniture making, is customization. In many homes, no effort was spared in fitting the details of the house to its occupants. The other theme is restraint, the prevalence of a minimalist approach, where form must stand the test of function, and less can be enough. Taken together, these two divergent themes produce personal, whimsical, comfortable places to live, homes which, surprisingly, cost less than their equivalent mass-produced counterparts.

One change of attitude that makes this possible is a marked reduction in the velocity of the homebuilders. Most independent homesteaders, when asked where they expect to be in ten years, answer, "Right here." For conventional residential housing, where seven years is the average time a family stays in a house, standardization is the watchword and customization a bugaboo: "You can't do that," says the contractor, "it will reduce the resale value." For successful independent homesteaders, resale is not a consideration, and their children quickly let me know that they too thought of these beloved new homesteads as their ancestral homes, which would be occupied by the family for generations.

I found evidence of loving customization everywhere I looked, in old houses as well as new, on-the-grid and off-the-grid. Energy-consciousness is clearly accompanied by a willingness to make a commitment to a home and invest ourselves in it. In some older houses, the original structure could be seen only as a wall here and a doorway there, a piece of brick hearth beneath the high-tech woodstove, as if new, more sensible layers had grown over the old. This approach was most natural in houses built before the debased values and techniques of commodity construction took over. I was heartened to find that fine old shelters could be adapted to work as energy-conscious homes. From within, through the eyes of the family, these were and are elegant machines for living.

Building the furniture, cabinets, and other details to fit their purpose makes the house more livable (as long as the habits of residents remain the same) and often allows for savings in space. Built-in furnishings are usually discouraged in commodity houses because one tenant's preferences may be quite different from the next, and accommodations are therefore reduced to low, common denominators—so that no one feels very comfortable, as one realtor expressed it.

For the most part, the graceful art of building to enhance life has fallen to the pressures of an impatient mass marketplace. And yet it is surprising how much time and effort can be saved if, for example, the kitchen matches the kind of foods prepared, and the people preparing them. Standard counter heights work well for standardized people, who seldom inhabit independent homes. Should we insist that a short or tall chef spend years of discomfort and resentment in a wrong-sized kitchen? The same goes for counters anywhere in the house, for the way our clothes and toilet articles and tools are stored, and for the treatment of all forms of work space and storage space and play space.

Independent homebuilders reinvent customized features with mixed results. In most cases the outcome is adequate, but, being reinvented, is not as well refined as solutions that have evolved. Much can be learned by looking at the work that has gone before, and the most successful customizers borrow from many cultures and disciplines. Where space is being optimized, customization is at its most essential, and inspiration can be sought in domains where personalization, spatial economy, and grace have been elevated to the level of high art: boatbuilding, Japanese homebuilding, fine furniture and woodworking, and luxury homes built before 1930. Although I hate to admit it, mobile homes and recreational vehicles are often triumphs of miniaturization; the execution may be trashy and the vehicle deplorable, but there are ideas there to be borrowed and humanized. Other good ideas may easily be found in books and magazines, and time spent in research can save you several cycles of building and rebuilding.

The refrigerator is one of the wildest cards in the home customization deck; because the trend is toward much better insulation, and no one is willing to give up interior space, the new box will always be too large. I saw many inventive solutions to this problem, most of them awkward. The only rational answer, if kitchen counters are to be built to accomodate the refrigerator, is to wildly overestimate its size in the original kitchen layout, and use the leftover space beside it as a broom closet, tray storage, or in some other easily modifiable and relocatable way.

Building into the Future

In the two decades since I closed in my house, dramatic changes in material availability and energy consciousness have transformed the way houses are built. While many of these changes are made in the interest of quicker construction, possibly at the expense of long-term durability, many of them have come about for reasons of resource efficiency and the development of new materials. Relatively fewer houses will be built in the next two decades, but the pressures to improve performance in key areas—heating and lighting, especially—are increasing, and so the pace of innovation will continue. Heads-up, forward-looking analysis on my part might have convinced me to push my local suppliers to provide me with better windows, for example, but they would still not have been as good or

The Brazil Effect

This term was originally coined, I believe, by Seymour Papert, an educational theorist, based on the coming of modern telephony to Brazil. Our telephone system has grown gradually and almost organically over nearly a century, and some of its far-flung branches are still examples of early technology: copper, relays, mechanical switches. Gradually, the fastest-growing parts of our phone system have been upgraded to fiber optics and solid-state switching, but repair workers in these areas must be trained in, and carry tools for, both technologies, and our system's performance is sometimes saddled with awkward interfaces between old and new.

In Brazil, the phone system was practically non-existent, and could therefore be built from scratch using the best available technologies. The repair people there only need to be trained in, and carry tools for, one technology. Brazil's telephone system costs less to build and maintain and works better than our own.

Is it wise, then, to wait until a technology matures before taking the first step? Of course not: far better to risk being cut by the leading edge of technology, to pay the price of being the first, and have the advantage of months or years of experience.

as inexpensive as windows retrofitted into my home today. This is called the Brazil effect; we can expect to be bitten and gifted by it over the life of anything we build.

A home exists in four dimensions, three of them in space and the fourth over time. A well-built home should outlive us, and serve our children well. Changes in the needs and abilities of a home's keepers and the new technologies available must feed the organic responses of a home. As long as it stands, we should stand prepared for change, so our home continues to shelter us well.

CHAPTER 11

GROWING WITH THE LAND

OUR HOMES GROW IN BEAUTY AS LIVING THINGS INSIDE AND OUTSIDE THEM grow, and as we work to care for this inimitable microcosm. The grounds about the home, the strand between our home place and the rest of the world, enfold us in an immediate source of food and oxygen. This surrounding microclimate profoundly affects the house's energy budget. A garden demonstrates our understanding of the wider bioregion, and is also a touchstone for the healthiness of our environmental adjustment. If we garden with the same arrogance with which we sometimes build our buildings, the garden further insults the natural world and precludes our fruitful relationship with the land.

Slash and burn is still a common approach to nature. Do we not owe it to our land to preserve its particular gifts? During construction, especially when contractors and heavy equipment are at large, we require a steward and champion with a clear vision of how the land is to be cared for. In most places, life is unbelievably regenerative, and plants may be drastically cut back without being killed, so long as their roots are not damaged by brutes with big tools. Before whacking the first bush or turning the first spadeful of earth it is worthwhile, even if difficult, to imagine the new home amidst its foliage, in order to adjust each to the other. The best adjustments foster and preserve life.

Trees are the land's leading citizens, and our relationship with them tells a story about our responses to the land. Trees often occupy the same settings we prefer for houses, and must be asked to yield, especially when space is limited. If trees must be felled to clear the homesite, we can reinvest their energy in the house directly by milling them and building with the lumber. The timbers in my house were milled by my own hand from windfalls, trash crop trees felled by a storm at the edge of the forest a mile east, with a primitive Alaskan mill, a chain saw with rollers. Made from commercially undesirable wood, these timbers are rough and too large for their purpose, and hanging them was a challenge, but they lend authority, solidity, and a powerful sense of place to my home.

Instead of building hastily, the forest moves incredibly slowly. Where I make my home, redwood trees are the oldest independent homebuilders, many having tended the same ground for more than two thousand years. This serene and stable neighborhood contains nothing more disorderly than a bluejay yanking near the stream. Redwoods grow fastest near flowing water, but the stream's patient undermining of their shallow underpinnings defeats these trees; when they fall, they release a long, narrow meadow for smaller lifeforms. On the hillslope a landslip, perhaps precipitated by human roadbuilding or the stream's more gradual work, shows how thin the soil is here: a hydroponic forest where the landscape holds together at the root level to resist time and weather. The oldest trees have patiently erected their massive structures and assembled their support communities on the flats between stream and slope. After centuries of accommodation to terrain and eroding forces, the remaining redwood forest is a harmonious settlement of ancient spirits.

The edge of the forest, the tree shore, is where life is most abundant. We may wish to select this place for our homesites, but it is wise to preserve as much of the original tree line as we can, so its existing community can enrich and naturalize the homestead. The precise relationship we choose to the tree line will be defined by climate, kind of trees, and perhaps the energy source we select.

Open space is desirable, and if our plot is entirely wooded, we will want to clear. Although the forest may present us with raw material, often the land that begs for release will not be in marketable trees but will have a young or trashy cover that will provide only firewood. If we are unfamiliar with the species of any large natives on our land, it is important to find someone who is.

Several landholders told me they were surprised by the vistas and features they discovered when they cleared their plot. At the very least, I would expect someone buying wooded land to climb the tallest available tree, or a tall ladder, to get a sense of the topography. Arranging a photographic overflight in a small plane provides an overview and baseline record that will be useful as long as the land is held. Two acres, especially if hilly, is a surprisingly large parcel, and can easily take a half day to explore thoroughly, but without thorough exploration, ideally conducted over several seasons and in different kinds of weather, how is it possible to find the best house site?

While thinking of vistas, consider "view-shed" and solar access as well. Unless we come to the remotest of locales, any changes we make, any clearing, building, or planting, may obtrude on our neighbors, their cherished views, and the way the sun falls upon them. It is one of the most unanswerable kinds of aggression when an unfriendly or heedless neighbor erects structures or plants trees that spoil our views, steal our sun, or impair the value of our holding. Just as we should consult with knowledgeable natives about the habits and suitability of trees, we should make forestry planning a neighborhood activity, working for a stable beauty that suits all.

Twenty thousand trees: Frank Dolan's story

Frank Dolan lives on one hundred acres of meadow and woodland at the end of a long dirt road high above California's Lost Coast.

There's a Chinese saying, the wise man chooses the place nobody else wants, so in the late sixties I thought I'd seek that. This is the first time I've been part of a community that isn't deteriorating, in fact it is still getting better twenty years later. My land requirements were simple: I needed reliable, gravity-fed water, good sun exposure, good soil, and good neighbors. And it had to satisfy my spiritual needs but be perceived by others as not attractive. I was looking for a place with a grim economic future, because I knew that I could always create a way to make a living. I prospected all over the west for two years, finally narrowing it down to southern Oregon and northern California, which had the temperate climate that fits my nature.

I started living here before there was a road, and so I packed everything in. The roads, as far as they went, were pretty primitive. It was sad to see the prairie oak go down a few years back: we used to tie off to it and winch cars up its hill. When I first drove through here in the early 1960s, the last of the old growth Doug fir logging was going on, and they had murdered the landscape. It looked worse than the pictures of Vietnam on TV. I didn't see how it could ever recover from the scalping they had given it, but in the early '70s, when I returned, it had recovered

Frank Dolan's thrifty Black Welsh Mountain sheep don't mind the rain; the home and parrot hospital are rigged to use PV, wind, and hydro power.

enough that you could see its potential. Twenty years after, it has almost healed (except there's little old growth, of course) because of our incredibly regenerative climate. You can still see the scars, many of which have been caused by post-logging, misguided, government fire suppression efforts. It's easy to see that the land is better off in private ownership, with individual landowners cherishing their own holding, rather than big chunks in the hands of corporations or national forest. In my twenty-two years here, I've planted nearly twenty thousand trees (of course not all survive); it wasn't that hard, and didn't cost that much. There are sequoias [redwoods] that I planted that are almost as big around as me and thirty feet tall.

The first business I tried here was an organic farm. In the third year, after successes with specialty peppers and eggplants, we grew twelve thousand eggplants on three-quarters of an acre. Closely followed by twelve thousand beetles, which taught us the limitations of mono-cropping. That attracted a good group of Pacific moles, for whom I've found the best remedy is a stick of Juicy Fruit gum down the active hole.

I got an early start with emus: for a pair that cost five hundred dollars, I was offered thirty-three thousand. Now there is a wonderful small flock of Black Welsh Mountain sheep on my steep upper pasture.

I still grow food, a good portion of what we consume, but for a cash crop eggplants have given way to hookbills: parrots, cockatoos, and other members of the Psittacine family, intelligent, opportunistic, and manipulative birds, well suited to being human companions. I pioneered a live diet of sprouted grains, automatic watering, and thermostatically controlled misting—an early masterwork of 12-volt spark-and-arc design—to create a humid rainforest atmosphere in the summer. I have long-lived, happy birds, but no more babies than other breeders. Like humans, hookbills breed in adversity, and often require stress to get them to reproduce. It's commonplace in the business to sell an unproductive pair, and have them breed soon after arrival at their new home; we've even had success crating a pair up and driving them to the city and back.

I'm moving from birds to bird hardware and supplies, because the birds make nonstop demands you can't ignore, while a mail-order business fits life here better. I've developed and we are readying a state-of-the-art incubator for market. And here's a trunk full of feathers, one of my by-products, that I must edit and send off to the Native American Church and to a project in Panama that supplies feathers so the natives won't have to kill wild birds.

When the house construction has reached a resting place (it's never finished), and there's enough stability to the business, I plan to get off the place and play.

The land here exerts a powerful selection process of its own, and we're lucky it does. Not long after I got here, I remember the morning our regional paper, the *San Francisco Chronicle,* ran the headline that made us famous: "Humboldt County's Outlaw Growers," with a front page map with an arrow pointing right at us. The article said you could get rich quick by moving to Garberville and growing dope. There followed exactly what I had tried to get away from: a four- or five- year boom. But by the end, the get-rich-quick guys gave up and got out, and left those the land selected, a tight community dedicated to the idea of taking care of ourselves, healing the land, and getting government out of our lives. The myth created by the media and the police, that this is a place of outlaws and dangers, spoiled the real estate market.

We are an endangered species. We started as beatniks—I did, in any case —then became hippies. People around here don't categorize, and there's a diverse cross section of styles, radical elements, wild west cowboys, new agers, computer programmers, organic farmers, and some that combine these opposites . . . Style is not important. We're eclectic, and spirit is at the center of everything. United Stand, which started as a building code resistance movement in Mendocino County, won its point and became inactive there, then swept into southern Humboldt. Everybody, of whatever style, agreed: Get out of my house, government! The land selected us all, a long time ago. When you have sanity and health why do you need government?

One of the things that made it possible for us to stay was the absence of the intrusion of government. We end up doing what should be everyone's right anyway: sheltering ourselves according to our individual needs. Fortunately, government leaves us alone for the most part. I have no building permit, and won't get one until it is a meaningful process rather than a method for social control and revenue generation. There are two conditions remaining in United Stand's dialogue with the county: A certification program where citizens (remember them?) can ask Planning and Building for advice on health and safety issues without being pursued for violations. The citizen pays for the services, but is not liable, and the inspectors are bound not to prosecute anything they see that doesn't conform. Secondly, there should be a grandfather program that lets us enter the permit process without paying fines. As long as the alternative owner-builder isn't respected, innovation will take place outside the system. "People should have the same freedom to choose their shelter as they have to choose their clothing (as long as it doesn't endanger the health and safety of someone else)." This is the credo of United Stand of Humboldt. The burden of proof should be on the government. The way it is now, if it doesn't specifically conform to what's gone before, it's illegal.

I assume that bureaucrats are essentially lazy. When they go on the attack, immediately ask them for documentation. That sets them back at least thirty days. Meanwhile, you get into their backyard, and find their dirty underwear. It's always there. My strategy is, only get angry when appropriate, and tell the media that you're doing it for free, to make the county a better place.

KMUD, our community public radio station, is a great resource and source of strength. For example, somebody will get on the air and tell folks about the current outrage the county's hatching: They want to rename rural roads. They'd destroy the history of the county. KMUD lets us generate the kind of energy that's required to back off the bureaucrats. It's the electronic link for our remote homes.

There's another Chinese saying, that a man who finishes his house,

A healthy lawn thrives on base soil from the composting privy.

dies. If I was forced to finish this house under time pressure from Big Brother, it would have been a very different, inferior house. I have to be able to solve problems in the time frame of my own life. I can't finish a part of the house until I can visualize it.

I have three composting toilets here of my own design, but the design doesn't really matter. To succeed with a compost toilet, you've got to get good at making compost in your garden first. Master the art of getting kitchen and garden scraps to compost (which is simply making microorganisms happy), and you'll be able to design a compost privy for your circumstances. My luxuriant green lawn out there used to be unfertile Laughlin-type soil, but I gradually covered it with the output of the privy, and encouraged grass to colonize it. The cuttings in turn are used in the privies and fertility increases as the years pass.

My greywater system took a couple of tries. The first one failed utterly; if I'd been doing it with Big Brother looking over my shoulder, he would have declared it a failure, and would have made me go back to the conventional model. I fine-tuned the idea, and now it works perfectly.

I do my electrical work with a pocketful of fuses; I call it the spark-and-arc method. I collect several springs into a 12,000-gallon ferro-concrete tank a third of a mile away, with 115 feet of head. I run two 2-inch plastic lines to the house, one from the top of the tank, and the other from the bottom, so I know if the tank is full or not. I chose 2-inch pipe because it was all off-the shelf, and quite a bit less expensive, but I have enough water most of the year to justify 3-inch and possibly a second generator. From Thanksgiving in a wet year until June, my two-nozzle hydro generator puts out 22 amps twenty-four hours a day; I know when a salamander has plugged one nozzle because the output falls to 7 amps. This rig has paid for itself many, many, many times over, even at PG&E's subsidized rates. But you can always use more power.

Garden Hazards

Where fire is a threat, our relationship to the forest or range must respect the way these ecosystems sustains themselves. Healthy dryland forests and grasslands thin and maintain themselves by tolerating small, periodic, lightning-set fires. Homes are not as tolerant as the rest of the flora to this periodic fire release, and often do not perform well even in the mildest grass fire. By choosing fire-resistant roofing and siding, and by replacing the immediate forest understory with plantings that do not burn well, we may be able to hedge on the fire fighters' bare earth requirement somewhat, but, as has been repeatedly learned in the Oakland-Berkeley hills, the relationship between homes and dryland forests is a precarious one. Besides careful planting, those who build in such a place must plan several emergency escape routes within the house and from the property.

Both sun and storm enter into any calculations about trees. Anyone who has heard a tree fall on a stormy night, or has seen the wreckage, will want to be sure that a climax forest, trees which have reached their maturity and have a propensity for falling, at least a tree-length away from the house and any important or valuable possessions. A car under a fallen tree is particularly forlorn, in that the fall could have been foreseen and the car moved. But a tree line may also offer, as it does on my land, the only defense against nagging winds. If trees are still growing, we must plan our solar exposure with their eventual size in mind, or plan someday to cut them. Where cutting is an aspect of forest management, and the windbreak is important to the home's microclimate, we must consciously plant replacement saplings well before the older trees are ready to harvest.

Deciduous trees and bushes shade us from too much heat in summer, but let in winter sun through leafless wintertime branches; seasonal shade suits our cooling requirements perfectly. In addition to being deciduous, most orchard trees offer up other seasonal delights: blossoms in spring and fruit in their turn, in late summer.

When planting trees, we may wish to look first among the regional flora. Gardening arrogance is at its worst when plantings are taken wholly out of our regional context, and require water, fertilizer, and protection to survive. Exotics are a good choice only if they come from a similar environment and can be controlled. Japanese black pines, a low, slow-growing species from the salt-blown headlands opposite my own, abide well with our native shore pines, and thrive without driving the native out, and so add a good counterpoint.

For a demonstration of bad species management, take a trip to Hawaii. Here in these most isolated of all land masses, tropical growing conditions and good soil have allowed imported flora and fauna to overpower what must once have been a wonderland of unique species. Generally for gain, but often just for sentiment or a misguided sense of fitness, exotic species were imported from every continent and have now forced out the native species, which never had to compete before. It is estimated that less than 20 percent of the original Hawaiian species survive, and native species cover less than 10 percent of the land area. Only recently have we become aware of the magnitude of this tragedy, as aggressive exotics wipe out native species from Florida to Alaska, from Hawaii to Maine. Introduction of an exotic species may begin a massive and irreversible environmental change which will affect every species in the ecosystem. Hawaii has been overgrown by the banyan, a large tree native to the Indian subcontinent that reproduces through tiny seeds which are eaten by birds and are carried as far as the birds fly. It is supposed that the banyan, which dominates Hawaii's humid forests by choking out dozens of more useful natives including sandalwood, came to the islands just a century ago as a houseplant.

Ark Maui: Wittt Billlman's story

At the precipice's edge on Maui's east shore, Wittt Billlman, musician and political theoretician (as you will see), lives in a simple, off-the-grid home. As he talked, whales breached below us.

As I get more serious about my endeavors, the more I am considered a comedian. Or a karmedian. You can spell karma with the H or without, depending on how Tibetan you are feeling. Spelling is a very important subject, so be sure you get my name right, triple T, triple L. I favor everyone adding a personalized twenty-seventh letter to the alphabet, so we can play with the jokers, too. Fifty-two letters doesn't seem to be enough in these times.

I came to Maui on a triple: I was looking for my high school math teacher, Jim Loomis, and I knew he was here; I was in a ten-piece Brazilian dance band then, playing a gig in Portland, Oregon, and a woman came up to me and said I was going to Maui, she could see it in my eyes; I still didn't take it seriously until I met another woman who said she was going to Maui, and asked me if I would come with her. It ended up all three of them lived on the same road, within a short distance of where I've been living nine days a week ever since (I bought the land from George Harrison, and I know it took him eight days a week to get it, and I'm sure I work longer hours). We're all from two different galaxies, you know: the Plaiedes . . . and the Work-days.

The first ten years I lived on candle-power. Water comes from rain and gravity flow. I've experienced duress, expense, and time keeping the utilities out of our neighborhood. My energy needs evolved, and I got one solar module, although I've recently been considering getting grid of it. It powers my electric piano, but I'm thinking about going acoustic, and using the power to dehumidify the baby grand instead. I like that image: a row of bulbs inside the piano keeping it warm and dry, and at night, when we need light, we open the top. I've written a new song for the Billary Allegory in the White House (isn't that just a primer coat?). Would you like to hear the song? I call it, "Inhale to the Chief."

Politics is very important to me. Right now I'm running from office, with the Party party, and have been making all the right moves: I have a political insultant, and a full-time champagne manager. Originally I had no platform, because I think everybody should have both feet on the ground (assuming they have two feet, of course. I sincerely don't mean to insult anybody). My supporters have pressured me to get a platform, so I've adopted one with a big hole in the middle, so I can stand on the ground, surrounded by platform.

When Hawaii becomes a non-nation, those are the offices I'm especially interested in running from. We hope that Hawaii sets such a prezident (that's with a Z, because we're near the end, and should be using more Zs than ever). We're working very hard on Ark-Maui. Did you know that 87 percent of the endangered species are in the Islands, 60 percent on Maui, and that 30 percent of America's endangered species live only on Maui? We are pleased that more people are starting to kNoah-bout the Ark.

Money is a symptom of centralization, and I believe we have to decentralize to survive. Since we went on the uranium standard, I've come to favor hundred dollar bills over ones, seeing as how both really cost the government two cents to print. Lots of bananas and papayas

Wittt Billlman on the headland near his home on Maui's windward side.

grow on our land, and that's how we survive . . . until lunch. We have a little salad garden, some sweet potatoes, so with a little fish, we don't have many food needs. Our taxes started at two dollars twenty years ago, then twenty, two hundred, three thousand! That's 15,000 percent inflation. We pasturized our land, dedicated it to agriculture, and the taxes roll back to one thousand in two years.

Except for that, the county doesn't bother us much. I was working on a house for a neighbor the only time I ever saw the building inspector. By coincidence, I was just gluing the last joint in the water system, and a big rain cloud was coming up over the mountain when he showed up. He got out of his truck and looked around, so I went over and said "May I help you?" He looked me over and said, "Some folks call you hippy, but I call you pioneer." He got back in his truck and drove off.

Slash and Burn Gardening

Starting in 1880 the inland redwood forest, coastal meadow, and shore pine community that is now the village of Caspar was scalped, flattened, settled, scalped again and replanted with fast growing cash trees, poisoned, and abandoned to the ground squirrels and a new wave of settlers. The land was scarred by careless use of the tools, machines, and chemicals humans have recently employed in their war with nature. In places, fragile topsoil built up over centuries was carelessly scraped away and allowed to wash into the ocean so that murdered redwood trees might be stacked there, or so that the tree murderers could site their flatland camps.

Plants and building styles transplanted from other climates survived, but did not thrive. The worst newcomers, the eucalyptus from Australia, were imported by an ignorant speculator who had been told Down Under that they were fast-growing trees that made timber. What he was not told, or was too hurried to understand, was that there are many different eucalypts; he got the fast ones which make miserable timber and are only marginally useful for firewood as their ribbon grain makes the wood unsplittable when dry. Early land-holders thought one pine is like another, so they planted mountain species. Cypress, another misfit exotic, is a Mediterranean native that here becomes brittle, sheds limbs, and finally falls after a fifth of the years it enjoys in its homeland. All these exotics, unhappy with coastal fog and rain, bolt like berserk weeds to find sufficient sun; in less than a decade they can be leggy and unstable, and fall to the buffeting winter winds. But these trees grow fast in their early years, lending a brief but green and reassuring pretense of healthiness, which is all the developer requires. As a consolation, the young trees are handsome, and all this downed timber feeds our fireplaces.

Only lately have we come to understand that other trees, the original natives, are more fit for the place, and require less from us and from the fragile soil. When I came to Caspar a quarter-century ago, I too planted mountain pine and cypress, and have lived here long enough to see them steal the soil's vigor and fall before they attained maturity. The California State Department of Beaches and Parks, in a rare burst of environmental heroism brought on by the drought and the resulting fire hazard among the dying exotics, has instituted a program of removing them and replanting with the original inhabitants, the shore pine. I hope other land owners can see the obvious success of this program in just a few years.

In two decades I have learned a great deal, and the land has been forgiving and resilient. A Sitka spruce, an exemplar of its resplendently upright, pointy-needled species at the far southern edge of a natural range that extends north to the arctic tree line, is pushing out the western yellow pines as if to say, *begone*, this is *my* place. In its shadow, the forest micro-flora are regaining a toehold. The deer have returned, and my roses are their favorite food despite my best explanations that the native flora is better for them.

Beneath the Trees

If we are lucky, and have trees, we may begin to develop the understory, the more human-scale accompaniment that plays at window level and below. Greenery softens harsh weather, humidifying and cooling the hottest days and slowing winter's gusts. Deciduous plants, especially vines trained to a trellis, extend living space and offer blessed shade in summer but retire graciously when the sun comes in more weakly at a lower angle to add welcome domestic warmth. Seasonal plantings draw our attention outside, and remind us that we are a part of a larger picture.

The synergy between house and garden is most important at this scale. Plants which grow happily as natives on a home site may be displaced when a house takes their sun or changes a feature of the microclimate on which they rely. Attention must be given to roof drainage, which can make existing places much wetter or dryer than they were before the house grew up. This same effect can be used to improve habitat. The shaded north side of the house offers protection for delicate species; used carefully, well-protected from the wind and selectively viewed through small windows, this can be used to lighten rooms on the utilitarian dark side of a house. By capturing and redirecting roof drainage, plants that usually require more or less than the local rain can be watered without waste. Roof catchment, a large agricultural cistern, and solar water pumping offer particularly appropriate and satisfying ways to lengthen growing seasons and encourage wetland plants in a normally dryer site. At one farm, a subarray of photovoltaic panels contributes to the home electrical supply in winter when there is less sun; when the rains stop, this array is switched over to pumping water from the river at the property's low point to a large tank on the ridge above for gravity irrigation.

A bird's eye view of the house, its approaches and outbuildings and its enfolding garden, reveals unexpected patterns that can be appreciated and amplified. Devote some thought to your expected traffic patterns, and undertake plantings and clearings accordingly. Repeated travel along a path can turn it into a mire during the wet season. I was surprised by how many independent homesteaders laughed this off: "Mud season's like this; it will dry up soon." Having lived in a rainy spot for a long time, I am impatient with this mud hero approach. Too much energy gets lost dealing with mud, stuck in the mud, muddy feet, muddy waters, for us to wait until it goes away. There are a few spots on the planet where mud is truly unavoidable; elsewhere, a modest but comprehensive program of pathways and driveways saves more energy than can be told. Where mud is bottomless and truckloads of gravel sink quickly from sight, may I suggest excluding vehicles from the area near the homestead during mud times, and providing a developed walkway (and a wheelbarrow for difficult loads)?

And when planning to add outbuildings, we must ask important questions about the impact on our present home place. Will the new structure improve wind shelter or will it create a wind funnel during bad weather? How will it change our favorite views or paths? Being careful not to fall, the best way for us ground dwellers to envision these changes is to take to the roof on a sunny day.

The Kitchen Garden

It is a poor home indeed where no fraction of the food we eat comes from our own garden. A good place to start is by visiting an experienced local botanist, who can tell what native plants are edible, and how best to encourage them in your garden. Neighborhood gardeners, too, will offer a wealth of specific knowledge about what grows well, and how favorites can be coddled. If you time your visit properly, you may come away with enough starts to begin your first year's experiment. (Avoid zucchini season for making these visits.)

Small gardens succeed when more grandiose plans fall of their own weight. If we aim to live on a plot for a long time, successful establishment of a single perennial should make us happy with a year's efforts. In a benign garden environment like Caspar's, the battle is slightly different from that in a harsher environment: whereas in Vermont the gardener will go out after the last heavy snowfall to see what favorite plants have been able to overwinter, I go out with jungle-cutting implements to see which of my favorites has not been choked out by spring's onslaught of riotous weeds. Wherever we plant with intention, we find our plans may not coincide with those of the greater gardener, who usually has her way. By starting small, with a few plants or one bed at a time, we can be heartened by the balance of success and failure. Nature abhors single crops and straight lines, and attacks them with all the pests and pestilence at her disposal. Independent homesteaders resist bringing chemical treatments near their homes, and particularly near their food crops. Once again, rejection of conventional petrochemical wisdom leads to rediscovery of traditional truths. The result is tastier tomatoes and peaches, hardier regional crops, and reduced load on the planet's energy resources. We find that traditional agricultural practices—companion planting, attentive care and watering, biological pest control, and a gracious respect for the whims of the greater gardener—are more satisfying and effective than aggressive, expensive technological interventions.

In the garden as in the house, the goal is independence. Homesteaders prefer systems that require minimal maintenance once established. By studying the home site we can learn about the garden, where our sensitivity to microclimate resolves to a focus on even finer features—rocks, trees, and shrubs—and a concern for the effect of person-sized and house-sized masses in nature. Over seasons of firm yet gentle urgings, we tune the garden to be our most perfect energy source, harvesting sunlight and providing for our physical and psychic needs. By discerning and learning to follow the natural rhythms and tendencies of the land, we further decrease our reliance on imported energy. A completely sustainable garden requires only what comes to it naturally—light and water—and produces no waste. Only nature does it perfectly, but we can do well.

Applying whole-life cost analysis to soil amendments, planting materials, and gardening hardware is just as reasonable here as in the house, and you will find that by this measure, use of some highly innovative technologies is economical. Drip irrigation, for instance, pinpoints water delivery, consuming less water and demanding less of our time; this method uses petrochemicals reasonably: instead of burning them up, they are embodied in a plastic irrigation system that will be useful for many years if well-planned. Cold frames, greenhouses, and other techniques for lengthening the growing season can easily be cost-justified in any productive garden.

The kitchen garden should be awarded the most prized solar exposure. Within the garden, sensitivity to the solar requirements and growing habits of a given crop will be repaid: low-to-the-ground eggplants and squash require as much light as they can be given, while lettuce prefers a smaller portion. To a home energy harvester, such considerations rapidly become intuitive.

Discussing home food production from a garden, Richard Gottlieb applies a ratio, like the *solar fraction* (which, as you may remember from chapter 8, represents the amount of electrical energy that can be reasonably expected to come from the sun). In the solar blue zone of Vermont, he says, it is very difficult to push the *local food fraction* above 10 percent homegrown food. Other growers in the region, employing season-lengthening techniques and hybrids bred to cope with local conditions, report results of up to 50 percent.

Growing Our Energy

Sustainable biofuels are nearly ready to replace some of our fossil fuel needs. Hydrogen and methane can be renewably harvested using the sun, and offer us an efficient way to store heat. These technologies are on the brink of practicality, and some pioneers are using them already.

The idea behind biofuel is that a suitable crop is grown for its energy content and its ability to improve (or at least not deplete) the soil. Fast growing, nitrogen-fixing crops like Gliricidia or fava beans are ideal candidates. Mature plants are chopped up, mixed with water, placed in a closed tank, and anaerobically digested by a community of microbes already in residence in the digester. Methane bubbles off to be captured, dried and desulfured, stored, and used as a fuel.

Hydrogen generation is simplicity itself: nudge water molecules with sufficient electricity, and out pop hydrogen and oxygen. In a curiously plantlike process, solar electricity is used to hydrolyze water, to break it into its gaseous components, oxygen and hydrogen. Hydrolysis liberates oxygen, which we can always use, and hydrogen, which is captured, stored, and used for heating tasks.

Hydrogen is a particularly desirable fuel. Our shared cultural image of

Wes and Linda Edward's furnace room: even in cold climates, a building's south wall can provide most wintertime heat.

the exploding Hindenburg notwithstanding, hydrogen is easy to store, and a far less aggressive explosive than most petroleum-based products. It burns cleanly, producing no chemical gotcha!s. Methane is not quite so well-behaved during storage or combustion, but is manageable on a small scale. Neither lends itself to efficient liquification, and so the problem of transportation remains a puzzle, because vehicles need a light, concentrated fuel source.

The Garden Within

With or without intention, the whole south side of a building is a solar collector. Using glass and thermal mass, we can keep living space within a comfortable range of temperatures. In many of the houses I visited, the space between the glass and the mass was used to bring the garden indoors. People who live or have lived in a greenhouse rhapsodize about the joys. Amory Lovins and his iguana, Robert Sardinsky and his midwinter carrots, Mike Reynolds's banana tree, all testify to the delights of the living room jungle.

While the concept is natural, the execution is challenging. Indoor agriculture introduces a broad new variety of pests, petals, and pollenizers, and requires energetic, attentive, and even inspired management to look good and do well in the compact terrarium of a house. Some jungle vegetation is generous with its spores and pollen, so an allergic gardener will choose tropical species carefully and introduce them cautiously into the indoor garden.

In cold weather sites, the greenhouse restores humidity to air that has been freeze-dried outside, with beneficent effects for the house's inhabitants. In temperate and humid climates the greenhouse's added moisture and odors can be less welcome, and good ventilation is important. At Rocky Mountain Institute, a PV module coupled to a fan provides air circulation when the sun shines, load matching source.

Growing space inside the home restores our connection with growing things. Growth—flower, fruit, and the annual cycle of reproduction—make us glad; as Robert Sardinsky says, "I've never met anyone who isn't moved by a garden." Watching Arnie and Maria Valdez in their greenhouse living room, pinching dead leaves and tending the tomatoes as they talked to me, I could see how naturally such space becomes an expression of home life.

CHAPTER 12

GENERALISTS AND SPECIALISTS

MANY OF US WHO CHOOSE THE PATH OF ENERGY INDEPENDENCE CAN OFFER plausible economic reasons, but at heart we follow a deeper reasoning. We are curious about how best to care for ourselves and our surroundings, and are eager for the challenge of building and managing our own life-sustaining systems. We imagine that some of the best rewards come from consciously negotiated arrangements with the genius loci of the place we inhabit. We believe we can improve on conventional ways of caring for the biosphere, the local ecosystem, and our family. And so we take up the fool's cudgel, and become jacks of all trades.

It normally takes a mob of specialists to develop a home site, from realtors, planners, and architects through finish carpenters, appliance installers, and landscapers. If all the tradespeople were to work together as a team on my house, with me their captain, all would benefit—but has this been your experience with the planning apparatus and the building trades? Amory Lovins, who has identified twenty-five development specialists, describes the process as a relay race, not a team effort: each specialist finishes his leg as fast (or as profitably) as possible, then passes the baton. The architect makes a design, seeking a dramatic approach, curb appeal, energy efficiency, whatever his design determinants happen to be for this job, then throws the blueprints over the wall to the mechanical engineer, saying, "Cool this!"—and so on, until a building is erected by expediency, greed, and inertia. But hold! This house is meant to shelter my family, keep us safe in times of trouble, let us laugh and make love. Our home, the domestic constellation of family, house, and land, is so close to the center of our lives that we ought to give nothing less than the best of our attention and the greatest part of our time and substance to its realization.

Renewable energy pioneers are often specialists who possess highly developed abilities to conduct research, amass knowledge, and develop technology. Some aspects of independent living promote high levels of special knowledge: we become energy devotees. Many of us (especially the men) are unworldly eggheads who are all too willing, even eager, to spend a thousand dollars on a solution that will save a nickel in a lifetime, because we like its technological elegance. We strive tenaciously to widen our vision, and are insistently curious about the edges of our specialties, from which we may peer over into other fields of endeavor.

However, where specialization tends to lump assumptions together and treat them as established fact in order to get to the product quickly, a more encompassing intelligence takes time to pick apart lumped assumptions, looking into the cracks between them for alternative solutions, missed details, and hidden opportunities.

Both tendencies are manifest in the minds and practices of independent homesteaders. Our earth-oriented, home-centered sensibility seems to border on animism, and pathetic fallacy pervades and informs our thinking, ceremoniously attributing consciousness and desire to photons and rocks, to rat snakes and people, to storms and to sunshine.

We find that we need to consider both the specialized particles and the generalized waves. Thinking about particles helps us understand our place in the midst of things, and waves tell us about grander currents and movements. Think, for instance, about particles of air: the warmer, more energetic ones want to rise, and the less energetic ones want to fall. We can take advantage of this by encouraging convection patterns inside our homes, by closing doors or opening drapes, by helping the particles do their work in a way that benefits us. Driving a nail through a piece of wood, we assist ourselves in the effort by visualizing the particles of wood that want to hold together while the nail wants to wedge them apart, and by visualizing the wood fibers parting to make a path for the nailpoint. Drive a nail hastily too near the cut end of a board, and as surely as particles have a finite ability to stick together, the wood splits and a nail, a piece of wood, and the time and energy it took to get them where they were meant to go is wasted. Phyllis Lindley, a veteran independent homesteader, explained, "We like hand tools for most work. It takes longer, but the mistakes are smaller and easier to fix." It may take longer to think about both the particles and the waves, but the results compensate.

We delight in tools. The cultural bias that determines which tools girls and boys pick up is directly short-circuited in a homestead where the women have a crucial role, for example, in maintaining the power system. Lafayette Young, master solar teacher and motorcycle mechanic, told of his pride when he found that his daughter knew without being told, and at quite a tender age, whether he needed a Phillips or a straight blade when he asked for a screwdriver. We make great gains for ourselves and for our children when we unwrap the mystery and work with the nuts and bolts of technology. We learn that brute force is seldom required, and that mind, intention, and the right tool will prevail. We learn that women can do anything men can do, and some other necessary things besides. We learn that two heads and four hands can lighten tasks, and that ears as well as mouths must be engaged for the synergy to take place. We learn the joy of cooperation, and the certitude that competition is rarely productive. We learn to take very good care of our fingers and our eyes, because we only get one set. All these little forms of learning combine to teach us to pay attention, to connect, to care just as passionately about the tiniest particles as we do about the planetary waves.

Popular electronic culture seems to be moving consumers toward a hands-off approach to the world. Lights turn on and off when one walks into a room; computers will reliably apprehend and transcribe our speech within the decade. Humans may, if we wish, choose to evolve out of our bodies. Independent homesteaders are torn between fascination with new technology and preference for a hands-on, do-it-myself manipulative and physical interaction with the world. Some of the mind-tools and electronic advantages we cannot imagine living without were barely conceivable two decades ago. Yet we rejoice in our bodies and their attachment to the earth, and readily take our values and rhythms from nature.

As human particles in a large and crowded Gaian organism, it is easy to disconnect from the brute force of grid-supplied electricity even as its suppliers would like to convince us that energy addiction is in our cultural bloodstream and distinguishes us from the animals. In scarcely a century, we have become completely assimilated into much broader and more appropriate electronic grids: telephone, radio, and television. Of the nearly one hundred homes I visited, two were without telephones, by choice of their owners. Every other home, no matter how remote, was connected to the world by phone: marine radio-telephone, over which only one person at a time can speak, is a primitive solution, but it gets the job done; duplex radio-telephones, over which both parties can talk at once, are much more costly; hot-wired portable telephones with directional antennae are practical if only quasi-legal. Somehow, the telephone network can be extended far enough to reach even those living beyond the fringes of the electrical grid, if only because a two-conductor wire snakes easily for miles through the woods.

Tele-presence, our emerging ability to conduct business through the telephone and its partners fax and modem—moving electrons instead of protoplasm, as Amory Lovins expresses it—offers many of us a chance to be as effective from the end of the road as we could be in any workplace. Independent homesteaders spend increasingly more time at home, in order to pursue our part-time energy hobbies. If we require sophisticated tools, we make them run on homemade energy. If our work must be done in a lab, we build a lab at home. If our work is collaborative, we expand our home and family to accommodate our co-workers. We find ways to work in one place, and communicate our work elsewhere electronically.

The conventional house is designed and built to tolerate neglect. For independent homesteaders, the home raises the stakes of going away, because it requires that certain attentions be paid to it, and may perform less than optimally when we leave it alone and commute to work. Energy self-sufficiency demands and rewards our attendance; those rewards and our yen for independence endow the home with a creative, familiar, protective, and supportive atmosphere, and the combination is hard to leave. Living independently runs so much at right angles to conventional employment that home workers are forcefully encouraged to do innovative work.

For some of us, a home-centered existence simplifies our lives and reduces our economic needs. Time is one big gain: during the time it takes to dress for work and commute, workers must often employ specialists to perform tasks they might prefer to do themselves—child care, food gathering and preparation, and energy management. Workers indulge themselves with retail therapy, spending money to compensate themselves for their lost time doing unpleasant work. Home workers avoid these and other wastes because the homestead becomes increasingly identified with entertainment and pleasure. We invest our savings in a home place, in improved energy equipment, in space that fits us better, in more time for selves and family.

Fortunately, we can spiral in toward a home-centered lifestyle gradually if we find it attractive. Several home workers told me that they first heard a faint call, and began a long, fascinating inquiry into their lives, habits, talents, and needs. The number of explorers grows, and the exploration grows easier, every day. Witness the widening consciousness about energy and global impact. We might begin with a few compact fluorescent bulbs, a recommitment to the family, a take-home work project that allows us to call in well, a do-it-yourself solar experiment, a recycling bin, or a compost bucket. At some point, escape velocity may be attained, and the independent adventure is underway. If not, some responsible practices have become habitual, and some of the other, older habits will inevitably be jettisoned.

For land and family: Wes and Linda Edwards's story

Linda and Wes Edwards live in an off-the-grid solar house on a heavily wooded ridge a couple of folds back from the ocean, near Ettersburg, in southern Humboldt County, California. Their daughter lives nearby. They spend as much time at home as they can; their land burgeons from their attention.

Wes: We were born and married in Rochester, New York. In the late sixties, we had a revelation, that we should be growing our own food, and having something to say about our environment. We thought we would like to build our own house, and if we were to move, we thought we might as well chose a better climate. We arrived in Pleasant Hill, near San Francisco, in 1970 with the idea that we'd like to move to the country, but when we told the realtors we wanted rural property, they laughed: it was very costly. So we started looking for property in our camper, mapped out California, and started our big adventure every weekend.

A friend of a friend had eighty acres in Humboldt County. We were trying to comprehend what eighty acres must feel like! We came up and looked at his house, and were taken by the area. It was a hot summer day, and he told us where to find the swimming hole. When we found it, there was a nude baseball game going on, and we met some nice people. We decided we'd found a wonderful place. We went into town and found a realtor, who really seemed to listen. We told him we needed lots of water, a southern exposure, trees for firewood, and a site to build on. He immediately showed us this land. We didn't want to say we loved it too fast, but we camped here that night, and left our deposit with the realtor on our way back through town the next day.

We camped here every weekend, and built a ten-by-ten cabin with lumber scrounged weekday nights from the dumpsters in the city, and hauled in a trailer behind our camper. We worked with the land all weekend, trying to figure out how to use it. The winter of '73 was a killer, one hundred twenty inches of rain; we're glad we didn't come up here to live until spring. We were awed by the land: it was so much more, and there was so much more of it, than we had ever imagined. We sold all our stuff, our appliances and conventional possessions, and ended up with all our remaining furnishings and belongings in a pile under a tarp. Our daughters were thirteen and fourteen when we moved.

They'd really been a big part of our decision to move, because things weren't good for them where we were. I remember really late one night, we'd just finished a new roof on our city house, and Linda and I were sitting up there, at two-thirty in the morning, and this car drives up, a door opens and a couple of beer cans roll out, and then this lout staggers up the walkway. "Can I help you?" I ask him from the roof, which doesn't seem to strike him as too strange, because he says, "Can Patty come out?"

Moving to the country was a philosophical decision for us: getting away from neighbors in our faces, where we couldn't be private with our space. The land came first, and it came without power. At the time I remember thinking, I can deal with that! We found out that power is a survival issue.

We fenced three and a half acres so we could have a garden, increased the size of our natural pond, started studying house plans and doing electrical things.

The Edwards home makes the most of the sun.

Linda: I hated kerosene fumes, and we couldn't read with kerosene lighting.

Wes: Our first power source was the car. We'd drive to town and back, which charged an extra set of batteries, then plug the house into the car when we got home. In the early seventies, there wasn't much equipment available for remote living. We got a wind machine, set it up on the ridge top, and moved batteries back and forth between the house and the wind machine.

Generator-based systems are expensive. The units are costly to begin with, and even with intensive (and expensive) maintenance, they have a finite life. Their real problem is that they consume fossil fuels at an alarming rate. I figured in 1986 that, all costs included, I was paying seventy-five cents a kilowatt-hour to run a generator. We only have to run ours in the fall, when it's cloudy and the hydro system hasn't started yet. We run it as little as possible, and when it's running we load it up with the battery charger, washer, dryer, and anything else we need to use. It's not an efficient way to charge the batteries.

When we first came here, our plan was to have a neighborhood school on the ridge, and we even poured the foundation. But our girls could already do what high school students needed to do . . .

Linda: And there was so much for them to learn here, homesteading, how to grow things, build things. It worked very well.

Wes: At first, the girls acted like we were going to a foreign country: "Dad, we're not going to meet any boys."

We lost Patty, our older daughter, in '76, to complications of cystic fibrosis. She was sixteen.

Linda: We took some time out then.

Wes: I was non-productive for several years. It felt like my heart was ripped out. That was a setback . . .

My true love is being out in the field, installing systems. PVs: I love to handle them. What a concept! Sun falls on them, and out comes electricity. I get to work with wonderful people, houses, and equipment. There's not a day goes by but I say, and I get paid for this? I'm getting more people who have sold suburban homes and are ready to build serious solar electric systems. In five years, they'll certainly be competitive with the utilities.

With utility power, you pay for convenience. If you run your own system, you must get involved with it, learn about meters and maintenance. And you must know and stay within the limits of your system. It's like *Zen and the Art of Motorcycle Maintenance,* you've got to get involved with the machine. Otherwise, you'll be calling somebody like me a lot. I like to see the batteries clean, and the terminals greased. Of course you can pay to have the system serviced, but that's not the idea.

Everybody uses more power than they've got. I make sure every one of my systems has good metering. Here, we live (and die) by the amp-hour meter. And education has to be part of every installation: How many hours can be used before you run the generator, how do you equalize the batteries. Batteries are the weakest link, and the least understood, and they get abused because they're not understood. Install good metering, and believe what it tells you!

In my experience, it's the women who know how to listen and learn. Maybe men have a block about listening to technical instruction. If I get a sense a man's not listening, I explain to the woman. I get calls all the time, and I can't believe what I hear. The guy says, "The meter's broken; I didn't use that much electricity, just a few lights." I talk to his wife, and she tells me the TV was on all day, and this and that, and he spent 900 amp-hours. Hmmm, not bad, 900 amp-hours out of a 720-amp-hour battery bank. I work with her so she understands the batteries, charging, equalizing and all, and now I'm confident it will all work right.

Wild west: Wayne and Debbie Robertson's story

Wayne and Debbie Robertson live with their two young children on a hill high above Willits, California. Both work in the alternative energy industry, but the part of their story that shines for me is their sense of family.

Wayne: We got tired of working all week so we could afford to get away on weekends from where we lived—why not live where we wanted to be? I'm an anarchist: it's a threat to have my energy come to me at someone else's pleasure. I see Igor with his hand on this big switch.

We produce the energy we need, and there's a beauty to that. If I had to pay a utility bill, I'd feel punished.

Debbie: If I'm missing any modern conveniences, I've been living off-the-grid so long, I don't know what they are . . .

Wayne: We have everything we need. Suppliers send me alternative energy equipment to test here, and I try to destroy everything I get. That helps the technology improve. A remote home is the toughest test there is for electrical equipment, because there's no pattern to the loads. When we were building, we ran four worm-drive saws off PV using a Trace inverter. My friends, who were building with me, loved working without the noise of a generator. We've got all the modern electronic kitchen equipment, including a microwave. You should see the surge when the dryer starts up! Maybe we could use a freezer, except we're not big carnivores.

Wayne and Debbie Robertson and their two children live in this comfortable energy-independent home above Willits, California.

Debbie: But there's all that tomato sauce in the refrigerator's freezer; it'd be nice to have a better place for that.

Wayne: Here on Third Gate Road, there are no power lines. On the other roads, the people that got plugged in have their freezers and refrigerators and other power suckers. If everybody generated their own power, they'd be acutely aware of power usage. We're more in tune with natural rhythms this way.

Debbie: Our kids were born out here, and they wonder why anyone wouldn't want to use free electricity.

Wayne: I remember walking with Sean in town when he was about five, and he pointed up to a power pole and said, "What's that, Dad?" I told him people get their electricity from it, and he asked, "You mean, most people don't use PV?"

Debbie: Sean and Lisa are great recyclers, always trying to figure out ways to reuse things instead of throwing them away. Their schoolmates are all off-the-gridders, too, and even the school was PV-powered until recently, so they've never known another way.

Wayne: They're used to outlaw housing with voodoo electricity. They are mostly aware of energy conservation, and usually turn off the power when they leave a room.

I expect this house will be finished when I die. I've got to have something to do with my hands, and my eyes are always seeing things that could work better or be better made. Before we start a project, we get all the books we can, and research it, and I visualize it to the end. Plans help us get started, and I talk to the County guys, but we change things as we go along, and always see better ways to go, and ways to go farther.

When we left the city, I took my retirement and put it into this house. We heat it on less than a cord of wood a year because it's so well insulated. I'm a perfectionist about foaming around the windows and sealing cracks and holes. I even foam around the outlet boxes, so this house is *tight.* I plan to grow old in this house.

It takes half an hour to drive to town, and so I'm aiming to move my work up here, and only go down the hill once or twice a week. I do all my work by phone anyway: with a laptop and a phone I can work anywhere.

Debbie: As long as the kids are here, one of us will probably commute daily for them or for work. There aren't many other kids around, so ours are real resourceful about finding things to do.

Wayne: When we aren't torturing them. [*The kids giggle; they don't look very intimidated.*] They've got the pond, an ever-evolving tree house, and the woods.

Debbie: After school they get plenty of time with other kids before we pick them up. As they get older, I can see that will become a problem; it already is for other families back here.

Wayne: We call this area the Wild West. People that live back here are real different from each other. The year-rounders have learned how to live here, but the springtime residents have no regard for the land. They just come to rip it off. The more people that come, the more full-timers there'll be, and the better it will be. Right now, the people back here don't even pay their road assessment, so I had to get on the Road Committee to get any work done. Up until recently, the road ate a car a year.

Independent Children

Nothing ties us to our home place more strongly than children. Nonetheless, independent homesteaders are not immune from the forces that necessitate two-income families, latch-key programs, and television-as-childcare.

I looked for, and expected to find, ways in which off-the-grid children differed from their plugged-in peers. I found, instead, that children living on independent homesteads or in families striving to control their energy use are no different from children who live in any supportive family environment. People who work with children will quickly understand that this is not to say they are normal children; the supportive family environment is not at all common in our working culture, where family life is submerged in the daily struggle to make ends meet, and the television set is the chief pacifier and source of intelligence.

Sex stereotypes are markedly less rigid off-the-grid. As I already noted, system management functions are performed by whichever family member is present when the system needs something, even if the initial designers are often the men. There seems to be excellent technology transfer, in the sense that all family members soon understand the basics of the system and are willing to tune it and their activities to fit the circumstances. What appears to be a remarkable mastery of energy topics is nothing more than environmental awareness; most of these children have spent their lives on "voodoo power," and it seems quite natural to them.

Many children reported to me that their friends like to visit them because they love the freedom of a natural setting, and because the household's focus on energy management is intriguing. I was also informed, in surprisingly strong terms, that it was at times difficult to return the visit, because on-the-grid friends live in wasteful ways that make energy-conscious children uncomfortable. When asked what is missing from their lives, most children were stumped. "Maybe I could play video games more . . . but no, I like to go outdoors, even when it's raining," one boy told me.

Making the home into a rich educational environment benefits both children and their parents, whether children are schooled at home or in town. Simple child-oriented features like light switches and kitchen workspaces low enough for short people to reach, so they may take care of their own needs and participate in family work, as well as small doors on children's rooms and child-sized furnishings, help children to understand that the home belongs to them.

Home schooling is an attractive alternative to many off-the-grid families, and is enthusiastically embraced by the younger children, aged twelve and below. Their favorite part is being able to stay home, living in the learning environment, instead of visiting it for a short, intense period after enduring, for some rural students, a crushingly long commute. With some of the time pressure removed from learning, lessons come about naturally rather than under pressure from curriculum and the added complexity of too many students and too few teachers. Certainly, for the younger children and adults who involved themselves, home schooling was enjoyable and productive. One benefit of home schooling in the early years is that it often gathers together a group of children, and allows one or more adults to remain at home for work. This results in what might be called tribe-building, which continues over into the lives of the families involved no matter what educational paths they follow as the children grow up. Small groups of like-minded souls who congregate to educate their children together can present a less programmatic, more serendipitous way of learning than is typical in a classroom. Problem solving is approached from different directions because the lay-teachers themselves come from differing educational backgrounds.

As children enter their teenage years, educational content becomes less important than social interaction, and almost every teenager I talked to had expressed, and been granted, a chance to go to a conventional school. The children agreed that for them the most important part of school took place before, between, and after classes, despite formal learning sessions which they deemed interesting but mostly irrelevant. They expressed a strong appreciation for the truth and grudging respect for adults who told it—and showed anger and contempt that adult-delivered education is so slow to wake up to the planet's real plight and so quick to perpetuate easy lies in lieu of more complicated reality. I was unable to interview a broad enough sample to be certain, but my impression is that children kept in a home-schooling environment well into adolescence might be brilliant, but were narrow and somewhat maladroit socially. I conclude that peer work is so important for children of this age that any benefits of home schooling, unless an unreasonably large cohort of children can be assembled, is overshadowed. At the same time, my decade of teaching leads me to conclude that most of the learning that goes on during the teenage years comes from caring parents in a richly educational home environment rather than from a traditional classroom. Independent parents tend to raise independent children who are immune to much of the emptiness and waste that plagues their peers. And children who have enjoyed the tribe-building of early home schooling seem to make a good transition to conventional school on an academic basis, even when they find their new classmates distractible, disruptive, and unfocussed. When the parents involved in the earlier home schooling continue to involve themselves actively in secondary education and their children's teenage lives, the family relationships, although certainly strained, held.

At some point, most children graduate from their independent home lifestyle, and seek the bright lights and consumptive practices of the city. One strong motivation, of course, is a wish to meet more people and experience the world. Even for children who have travelled during their childhood, a desire to broaden their acquaintanceship and polish their abilities to exist in the greater world is noticeable, and not at all unlike the experience of children growing up in small communities everywhere. This causes temporary alarm among the parents, who may find that their only solace is redirecting their attention to enlarging the doors of children's rooms, and finding new uses for the space. As I say, this alarm is nearly always temporary, because most of the children soon report home about the shallowness and pointlessness of urban life, and move toward a more peaceful lifestyle, often near their parents, or to a more directed life in the city. Amazingly, they grow up, just as we did! It will be interesting to watch this process as it completes a second generation and enters its third.

A problem universally acknowledged by parents and children alike, and one which is obviously not a circumstance limited to independent-home-raised children, is the difficult passage when boys, beer, and internal combustion engines flow and churn together. In Southern Humboldt, where the off-the-grid lifestyle is the norm rather than the exception, the too-much-testosterone (TMT) problem is compounded by distance, bad weather, and difficult roads. Neither permissiveness nor strictness seem to work well, and the young men literally bounce off the walls, leaving tire tracks up the road's nearly vertical banks along the road between Redway and Briceland. It is no comfort that this problem appears to plague both on-the-grid and urban communities.

Southern Humboldt's community life is stable and benign, having been a haven for alternative lifestyles for two decades. The adults cherish their own childlike tendencies and many teenagers have been home-schooled and alternatively educated, so a tribal consciousness prevails: most adults and the louts themselves agree that we are all each other's parents and children. Society revolves around the Mateel Center, a community center that books world-class bands and that welcomes children, even these young bucks.

I can recommend only one stratagem for helping a young man survive the awkward years between the onset of personal awareness and the time he can take up some real personal power. Cautiously and with suitable adult aloofness, help him build, occupy, and maintain his own separate shelter—an independent homestead of his own.

Garbage

We are a nation of garbage specialists: we make garbage better than we do almost anything else. Taking proper care of our discard is an inescapable part of taking care of ourselves. Garbage, like the dependent home, is a modern invention, and a phenomenon without parallel in nature. Only upstart humankind has perfected the art of making waste so disgusting that it cannot serve as food for other organisms. We tell our children, with wonder in our voices, that the plains Indians used every part of the Buffalo. Of course they did! Every part was useful. In one century, a garbage culture has emerged that has the power to exhaust Earth's plenitude and bury the planet in pollution and waste. When our grandchildren take us to

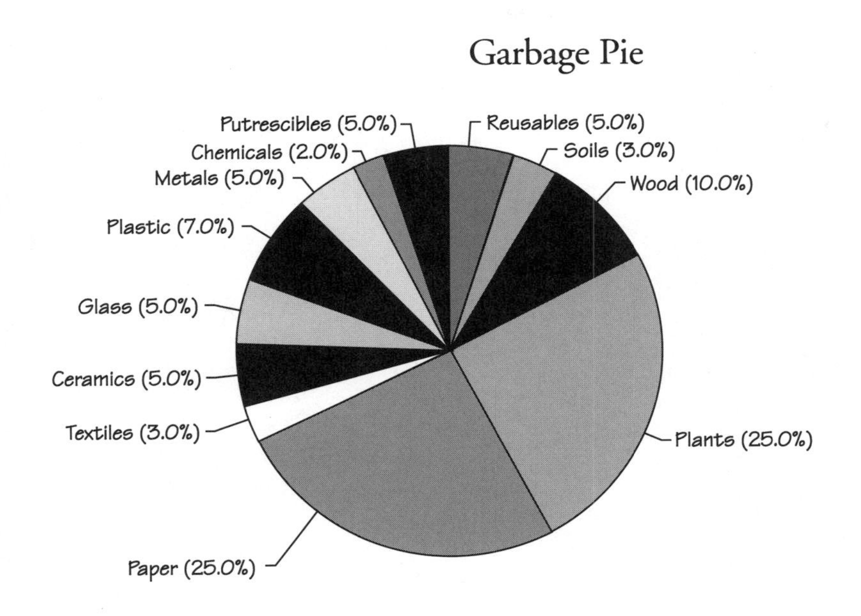

Figure 40. If different forms of waste are kept separate, they can be managed better. Here are the basic types of trash, and how these types add up at the Berkeley, California, dump.

task a few decades hence, we want to be able to say, "Not us; we were pioneers, trying to figure out how to do it right!" As each community's Mount Trashmore looms, market pressure—the cost of going to the dump—has made us more careful about our garbage. It is more expensive to dump now, and we pay less if we recycle. As a people, we are slowly beginning to get control of our trash habits, to reduce, reuse, and recycle because it makes sense.

Anyone who has tried to find something mistakenly thrown in the garbage understands the problem; the items in the trash share only one trait, that someone did not think they were of immediate use. This unilateral thought-lapse contributes to a motley mixture whose constituents are characterized primarily by their randomness. We compound mindless reflex with a fetish: once devalued by joining the trash, a discarded item is somehow sullied, and it is considered indecent to salvage it; polite people simply don't go through the trash, their own or anyone else's. So the reflex to trash something is irrevocable.

Our garbage waste problem and our energy waste problem are close kin; easily 75 percent of what we throw away can be turned to better use on the independent homestead. Start out by not buying trash or not bringing trash home. Then, for those of us who practice unlumping, it is a natural next step to separate and collect discards appropriately. Finally, by reusing the materials we stockpile, or by passing them along to industries that can profitably reclaim their precious elements, we further the reform of attitudes about the reclamation and redemption of garbage.

CHAPTER 13

LIVING WITH CONSTANT CHANGE

People seek equilibrium, but life is change. Managing an independent home requires almost continual attention to the cycles which abound in nature, followed, where warranted, by intelligent adjustments to the systems. Long-time off-the-gridders make these changes instinctively. We welcome the invitation to synchronize with our environment.

In chapter 4, I discussed the need for daily reconciliation of energy budget expectations and actualities. Each self-sufficient system—water, hot water, waste, space conditioning (heating and cooling), as well as electrical systems—requires such an audit, always with an eye to anticipating the next cycle, and managing it better. As this accounting becomes intuitive and automatic, and we become better managers of our living machine's adjustment to its environment, we might expect the process to take less time; maybe not surprisingly, we may well take more time, but also more pleasure, in maintenance chores. Independent homesteaders report that the mainspring of their days is wound by their house's performance. We start with a better home and better energy management, and we are always striving for improvements.

The shortest cycle, the passage of day and night, dominates the way a home performs. We gather energy during the day if we use photovoltaics.

Even wind has its diurnal rhythms. In an independent home, certain tasks (including bathing, laundry, and use of power tools) are arranged to fit the daily energy-gathering cycle. The next shortest cycle is comprised of the weather. In this realm, the patterns are never as sharply delineated as day and night, but every place has its generalities. In Caspar, in summertime, two or three pleasant days are often followed by three or four days of fog. In winter, storms seldom last more than three or four days. The satellite photo on the evening news helps us see and plan for the times when a line of two or three successive storms marches in on us from the Gulf of Alaska; if we rely on the sun for hot water, we bathe before the first storm hits.

The seasons impose another order on our weather. Seasonal patterns dominate our shorter-term weather forecasts, and modulate the amount of light that reaches us. Growing things respond to the seasons, as does falling water, wind, and outdoor temperature. Following along with these perennial rhythms is the annual round of holidays, birthdays, and anniversaries which figure importantly in the life of a home.

Almost everything that exists in time is cyclic, and so the house itself has a cycle, from inception to decay. First, as it is planned and built, the envisioned home slowly materializes. New houses, and any new structure, require a shake-down period during which we encounter and resolve unforeseen problems. Once the worst of the bugs are worked out, and the space becomes livable, we enter a stable phase. Independent homeowners are famous for the way they attenuate this stability; the bugs are scarcely dispatched when we find new ways to enlarge, improve, refine, and rebuild the space we have wrought. Inevitably, decay begins, and parts of the house are worn and abraded by the friction of life and seasons. Maintenance and repair are crucial here, or the cycle begins to wind inexorably down to the last phase, abandonment. In many independent homes, our insistence that we are never finished muddies this cycle, and we find ourselves coping with planning for new construction while also working to stave off decay. Within this century-long house cycle, we encounter the shorter cycles of equipment life, and the complicated decisions associated with repair or replacement.

Solar Cycles

When we take control of our own energy systems, we accept the daily responsibility of anticipating needs and opportunities, and readjusting our systems accordingly. Here is a simple example: if we know that a few hours of sun remain before a storm, we might check our water and power supplies; if they contain more than needed to weather the storm, it may be wash day, since we know that water consumed will soon be replaced and that power consumed may be replaced before night falls or the storm closes in. If we hesitate, the only thing conserved will be dirty laundry. We learn to be quick to seize opportunity.

Considerations of this kind are unique to independent systems. The same forces are at play in a conventional home, but responsibility for them has been tacitly delegated to invisible others, the water company and the electric utility. In our independent home, the systems and resources on which we rely are practically family members, and we consult them, checking their preferences, as we make our plans.

The annual modulation of daylight, as winter's brevity of days and lengthy nights reverses itself in spring and then reasserts itself in autumn, affects the way we dress, our daytime activities, even the hour and circumstances of our rising, dining, and going to bed. Each season has its homesteading tasks. In spring, we check and repair the ravages of winter, remove the storm-battenings, prepare the soil and put in our garden, and plan the summertime renovations that will make next winter easier. In summer we cherish our shade, tend our plantings, watch our batteries that they do not overcharge and our water that it is not overdrawn, make the changes the next winter will require and plan the changes we will put in place next spring to make summer more enjoyable. In fall we harvest and prepare for the onset of winter; we get in the firewood (ideally for the winter after next), check the chimney, confirm the health of batteries and equipment that will be called upon to give us comfort through winter's long nights, and begin to dream of spring. Wintertime's arduous totality,

Figure 41.

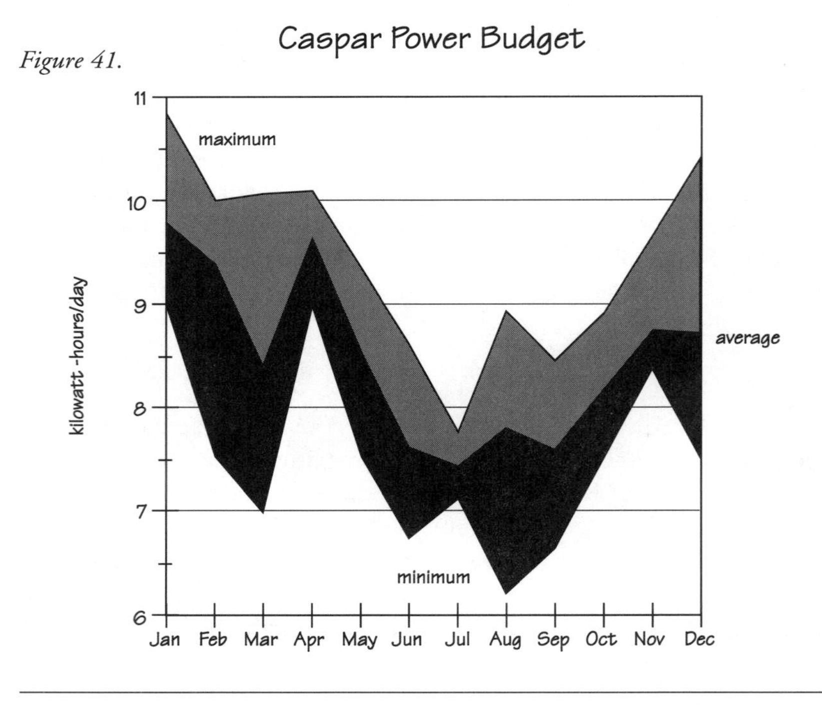

for all but those who live in the deserts and tropics, is the true test of an independent homestead: if it stays habitable through this, the rest of the year is easy. We watch the sky and the satellite photos, fill the woodbox when we sense a storm, batten down and hope as the storm batters past, then mop up and make the immediate repairs that cannot be postponed until spring. Those of us who choose to live at the end of the road are subject to the vagaries and blessings of sun and storm, and we come to love their endless pageant.

An interesting way of looking at our energy cycles is in the form of daily and seasonal source/demand curves. As variable as the days yet as immutable as site, structure, and family disposition dictate, this is a tool to use in understanding how to optimize our energy.

Generations

I find myself, at times, sitting on a step, a stool, a rock, contemplating a part of my house. When I built it, I knew that parts of it were right, absolutely; the foundation and the anchoring of frame to foundation had to be right, for instance, for I planned a house taller than its longest ground-level dimension, a sort of tree house in an environment where tree-levelling winds are a wintertime commonplace. The kitchen, a great favorite of mine, came right into focus very soon, even though some of its components are temporary, waiting for better replacements or subject to the designed-obsolescence that curses our consumer society. Some aspects, I knew, were close and would come closer by stepwise refinement over time, and could not be hurried. This is a notion with which my partner Rochelle has troubles. "Why can't we just do it right now?" she wonders. I reply, "Because I don't know what *right* is yet; I am not yet at the part in the cycle where I know the right way to proceed."

From the start I have known that parts of this house are temporary; funny, how some of those temporary parts persist!—the entry, or the stairway in the northwest corner I have always wanted to replace with a rock tower to anchor the whole house against the blasting northwesterlies. Gradually, as I live with this assemblage of permanent and temporary parts, new visions come to me, more or less grandiose or practical. Occasionally one will be so right that in a flurry of construction the vision becomes real. This is the pattern of a stubborn, naive builder inventing his own functional space. And so, as surely as salty wind peels paint, the unfinished details of the temporary sections decay, and must be patched, while just-finished additions prepare to face their first shake-down winter.

Some people may choose to immerse themselves in other pursuits, and let experts envision and construct their living space for them. The most visionary experts understand that homebuilding is a participatory activity that relies for its impetus and personality on the inspiration and insight provided by the homeowners. Pioneers may choose to go toe-to-toe with the forces and frustrations of living in space they have built, and settlers may find ways to have others tackle the details of homebuilding; either way, participation in the building process is undoubtedly the finest part of living with an independent home.

Families have cycles. All too quickly we go from the year when we toddler-proof the paint cabinet to the year when we yield up the family car's keys to our new-wheeled teenager. Residences are often planned and built at the time when family life centers around young children, who seemingly will always be with us. Rather quickly, especially in the slow paced time-frame of an owner-built house, children begin to leave. Small, dormitory-style accommodations work well for young children, but as they grow up, no space within our home will suffice. When that time comes, it works well to build, or help the children build, their own separate cabins. These stand-alone shelters make excellent test-beds for experimental tech-nology, fine guest cabins, special purpose outbuildings (Linda Edwards packages her mustard in her daughter's casita), and habitations for the children when they return on visits. Frank Lloyd Wright's solution to this conundrum was to delineate rooms for transients in a semi-permanent way: in a dwelling he built for a university professor's family, the children's rooms adjoined the living room; as children left and the need for larger public space increased, the bedroom walls were removed and the living room enlarged.

Haste in designing a house, and failure to plan for the future, creates space that will not take good care of us as our needs change. Architects and house designers have masterful ideas about stairway management and bathroom configuration, but they may forget that for the family this home is more than a short-term habitation, like a motel room. Homes that grow to match the needs of the families that live in them may not be as architecturally gracious, but they respond beautifully to the family's needs. When in the throes of inspiration about redesigning or adding on to a house, remember the long-cycle needs of the family. For example, how will our homes nurture us as we age? My house, with its three levels and lots of windows, will work fairly well for us as long as we can climb steps; I hope we will always be able to do that. Catwalks and belaying points around the windows will make it possible for me to maintain them up to a ripe old age, but at some point I will need to enlist someone more agile and foolish to work on the third floor windows and the roof. The electrical, plumbing, and heating systems are left in the open to the extent that the building codes allow, for ease of service, and I have tried to map and label their invisible segments. I would like to say that all circuits are labeled and their routing known; only in researching this book did it come clear to me how important this home documentation can be. Already, in a mere twenty years, I have completely lost two wire runs. I know where they start, but I've forgotten where they go. Forgetfulness and aging are inescapable conditions of the family's cycle in a home. Many independent homes are in difficult terrain, and are plagued with uneven floors, small doorways, and many other features that make them young people's houses. It will be a sad day when these imperfections force their once-proud owners to leave home forever.

Public Consciousness and Cyclic Change

When I started compiling a comprehensive list of cycles that homes live through, I missed a very important one, but Mac Rood reminded me. "Don't make long-term business decisions based on short-term government policies," he admonished. In 1993, after a mini-ice age in the political fortunes of solar energy, the U.S.'s second longest political regime of the twentieth century came to an end. With a new president from the alternative energy generation and an avowedly pro-environment vice president, we could hope for a return to the sympathetic Carter era, with fresh technology: A solar White House! Net Metering! And a rational, equitable, and forward-looking revisitation of the energy policies that begat PURPA! Cully Judd, who lives in an off-the-grid demonstration home in the middle of on-the-grid Honolulu, says "Hawaiian Electric Company should get down on their knees and *beg* to pay me a dollar for every one of my nonpolluting, absolutely renewable kilowatts!"

The solar industry has survived the nuclear winter, and is stronger for the experience. There are multiple cycles at play, economic, generational, the quick-time quadrennial political cycle and its weird perturbations for two-year congressmen and six-year senators. As an undercurrent to all else, there is the environmental cycle. Some are so deafened by the foreground noise of jobs vs. owls that the sounds of the natural world are inaudible, but for others these are an imperative heartbeat. Fickle fashion and public attention ride the crest of such cycles, and like all short cycles, the ride can be brutally fast, and end beak down in the rocks. Of the millions of active solar hot water systems, installed with tax credit money (an enlightened program) and cobbled together by fly-by-nights (a horrible outcome), how many are still in service? When they broke down, how many owners fixed them? No one really knows, for here was a program without follow-through, a dream scam for exploiters, and a black eye for solar technologies.

The relevant lesson here comes in the form of a cautionary tale from Amory Lovins: "Nixon made a wonderful speech about population, perhaps one of the best ever made by an American President . . . but nobody ever thanked him, and so he never mentioned it again." We can affect

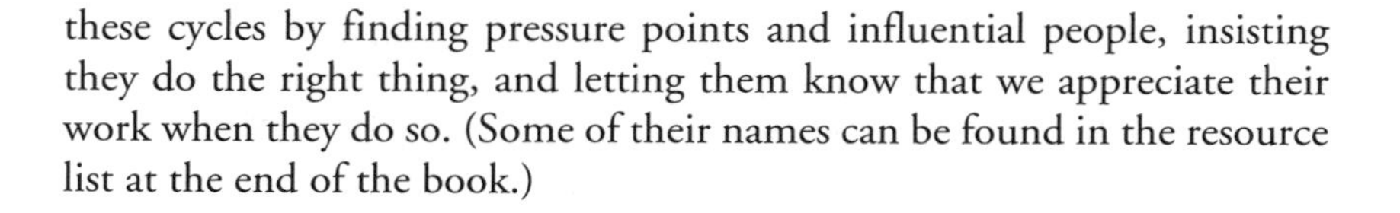

these cycles by finding pressure points and influential people, insisting they do the right thing, and letting them know that we appreciate their work when they do so. (Some of their names can be found in the resource list at the end of the book.)

Long-term plans: Mac Rood's story

Macrae (Mac) Rood directs Yestermorrow Design/Build School in Warren, Vermont. Yestermorrow offers classes that help people design and build their own homes. In his work, he has become acutely aware of the many cycles and fashions that affect the way we make investments, build homes, and live our lives.

Mac Rood.

Since 1980 we've been teaching people to take charge of their built environment by designing their own homes. We have seen some amazing changes in the basic assumptions in that time. Vermont has building codes, but we take them as possibilities, not constraints. Even if our students don't go on to nail their own houses together, we have produced informed clients. Warren is a hotbed of alternative architects, and I venture there are more architects per capita here than anywhere else in the world. Prickly Mountain, where I'm taking you, was a learning ground for some great ideas fifteen years ago.

We hold classes in framing, wiring, all the skills needed to plan and build for yourself. Our students come from all over. At first, graduates told us their architects didn't want to listen to their ideas, but, I'm glad to say, that's changing.

I just happened into this valley fresh out of college. A couple of days later, I was working with David Sellers, doing carpentry, then drafting. I never left. When I was sure I wanted to design and build around here, I went to Berkeley to get my architect credentials, then returned.

I prospected for hydro sites with Bill McDonough during the early PURPA days, when Jimmy Carter had us looking for energy anywhere we could. We looked at five hundred sites, and developed four; one of them, in Barnet, had ninety feet of drop in seventy feet of river, and produced 550 kilowatts. The deal was, the utility companies had to buy locally produced electricity at the avoided cost, in other words, the amount it would cost to build the power-generating facilities to provide that amount of power. We figured that, at Arab Oil Embargo rates, we could make enough to justify the investment. We wondered why business people failed to flock to invest with us. We were young, naive, and had the law on our side; obviously oil was going to cost more and more, and electricity would be in greater demand, so the government program would go on forever. We learned: what makes sense, and what governments do, are not necessarily the same for very long, so don't make long-term business decisions based on short-term government policies. We got the rate of return we expected one time, the first month we went on line with our first site, 7.8 cents a kilowatt hour, I think, and then the avoided cost started nosediving, 7.2 cents the next month, down now to 4 cents. Our best year, we made ten thousand dollars from the Warren site. We sold three sites, but I kept one, 50 kilowatts here in Warren. It's still a good idea.

The Mad River here in Warren has a good flow year-round. In winter, after a storm, it pounds through here so hard, the ground shakes. In this part of Vermont, the amount of water is right in proportion with the watershed area, and we drain twenty-seven square miles. This was an old mill site, but we had to rebuild the crib dam, headrace, and penstock. The penstock is four feet in diameter, eighty feet long, with twenty feet of head. We salvaged the turbine, but had a big induction motor rewound. It requires power from the grid to excite the coils, and it then produces synchronized power, so it can't fail and electrocute anyone. The insurance companies, of course, want outrageous sums to indemnify the power companies, and that part of the deal was and is horrible.

If we were to try to build today, we would go through a much more rigorous environmental review, and we might need fish ladders and what-all. Vermont's towns grew up around hydro sites and used the river as the sewer. Many of New England's rivers have been dead for decades, from near the headwaters to the sea, including this river. That doesn't mean the rivers need to remain dead, and we have cleaned them up impressively.

I made a mistake with this system, because now I see it would provide a perfect and very appropriate energy source for the house I built right above it. When I was building, the utility intertie looked very attractive, because fuel oil was expensive, and I could not imagine using 50 kilowatts of high-grade energy for heating a house. In retrospect, even with cheaper oil, it makes sense to "waste" the electricity for domestic heat, and share my excess with my immediate neighbors.

The Political Weather

We may attempt to shift our lives to take advantage of political cycles, but we may also try to guide those cycles. Politics blows hot and cold, first welcoming, then rejecting, then again welcoming alternative ideas about energy. At the state and municipal level, the agencies and commissions that regulate the activities of energy monopolies are meant to be guardians, insuring that the best possible local policies are instituted for consumers. Too often, these institutions attempt to maintain equilibrium by denying or resisting change.

Changes in energy sources are simply too strong to be denied. As Dennis Weaver, an advocate of energy sanity and an earthship dweller, says, "If we continue to stick our heads in the sand and ignore it, live in denial, and say we don't have a problem, it's only going to get worse." The work of regulating power monopolies is complicated by energy's technical nature and the fact that regulatory agencies are part of the political apparatus, administered by appointees, and infested by interested parties. Be that as it may, the public utility commissions are—or can be, because of their statutory obligation to serve the interests of the people, and their genuine power over the utilities—a powerful tool for reforming energy practices in a meaningful way. Increasingly, energy activists are learning the language, joining the club, and putting this tool to work.

Making cycles serve us: Leigh Seddon's story

Leigh Seddon designs hybrid renewable energy systems in Montpelier, the capital city of Vermont, where he lives on-the-grid in an energy-efficient house. His technical work has provided him with the necessary knowledge to intervene in the regulatory process, where he is in the forefront of a movement toward rational pricing and sane use of energy.

I make a basic assumption: People should expect energy suppliers to promote efficiency and renewable energy. This is the direction regulation is moving with the public utilities. Once we get that rationalized, then we can start working on the unregulated sector, like the petrochemical industry.

We certainly make efficiency the basis for our work. Architects bring energy problems to us. We prefer to work for the architect, so that we can design a transparent energy system for the client. If we do that properly, the owners should never know they're in an alternatively powered house, if they don't want to.

A decade ago, solar technology was cranky, and we just could not do that. Every system had to be sized uniquely for site and loads, and tinkered with interminably. We have enough experience, now, that we don't have to design every system anew. After all (I think this is Steve Baer's example), you don't hire an automobile designer, and the machin-

Leigh Seddon explains photovoltaic water pumping to students in Yemen.

ists to build every part, when you need a car. We do pre-engineered assemblies in our shop, then send the unit to the site. I like systems where the on-site work is simple, straightforward; the result is a trouble-free installation by local contractors. You get savings in design and installation, and system reliability goes way up.

We do photovoltaic (PV) systems, and solar hot water heating as well, and any other approach to renewable power use we can devise. For example, we're working on a grid-connected system where inexpensive off-peak energy is used to recharge a battery bank, which then powers heavy motor loads during the hours that electricity is in peak demand. In Vermont, commercial buildings are on time-of-day rates, and a peak kilowatt costs more than four times an off-peak kilowatt, seventeen cents in 1993. Even if you give up 20 percent efficiency in battery charge and discharge cycles, you can still get cheaper energy. Does it amortize the equipment? We can't tell yet. If our goal is a solar chicken in every pot, a home renewable energy system everybody can afford, we will need some angels to pioneer along the way.

Noel Perrin is one such pioneer who has a utility interconnect system. The story goes, he was pestered to do something about pollution by his environmental studies students. He was already connected to power lines, so it made sense to him to put PV modules on the roof, and sell his excess output to Central Vermont Public Service (CVPS). His utility contract was the first in Vermont. He gets four cents per kilowatt-hour and buys power back at about twice that. I guess it was a matter of principle, not economics, for Noel. At first CVPS wanted a million-dollar insurance policy and a full-time operator for a plant generating at best 200 kilowatt-hours a month! To shorten a seven month story, we got them to be a bit more realistic. Noel was able to satisfy them with a good homeowners policy and an inverter that shuts down if the grid fails. He's an example of the crusading angel, who establishes a principle at some personal cost, so others may follow. Now, it takes half an hour to set up another utility buy-back contract.

Figure 42. Time of Day metering charges a lower rate for electricity used during off-peak. This house system "time shifts" by recharging its batteries during off-peak times.

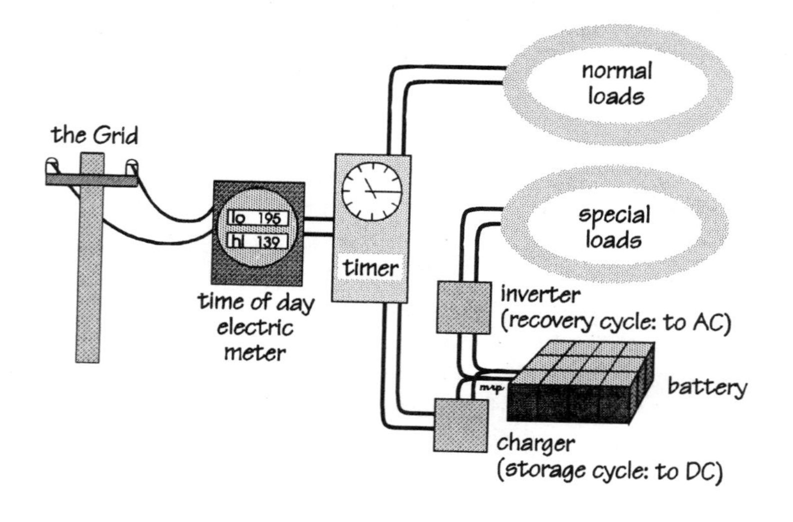

The key thing holding the renewable energy market back, I say, is a dysfunctional market, where we willingly spend billions for a war to keep oil prices low. What is the real cost of power? We don't have even a vague clue, but we know the subsidy is immense. In Vermont, and a few other states, there's a move to quantify the external costs of commercially generating electricity (we insiders call it the Externalities Docket, to keep people confused). For example, our utilities figure new coal capacity will cost eight cents a kilowatt-hour, but by the time you figure acid rain, air pollution, and other immediately identifiable environmental concerns, you can add seventeen cents more. Who knows what other costs will have to be paid down the road? Right now, Vermont regulators start much, much lower, with a 5 percent environmental adder, and another 10 percent for risks. Massachusetts adds 30 to 40 percent. Nevertheless, I am confident we'll have the obvious costs rationalized within five years.

In New England and California, we let our utilities earn more if they work to decrease consumption. We consider a compact fluorescent in a house to be a conservation power plant. In Vermont, the average electric customer uses 700 kilowatt-hours per month. We have worked with families and designed houses to get by on 100 kilowatt-hours quite comfortably. This turns out to be a very powerful tool for getting the utility's attention, which, we hope, spurs efforts towards super-insulation, super-refrigeration, solar hot water heating, PV systems, and other new technologies.

In five to ten years, I expect utility and PV costs to cross paths, and that's well within the life of a system, so I think it's already a viable investment. Assuming, of course, that we play fair with external costs and buy-back rates, and I think we will. Vermont is already considering a law to mandate utility buy-back rates at parity with their retail prices.

Earth Cycles and Preparedness

In the context of geological time, where the scale is grand, on the order of millennia, energy flows change direction often. It is not unusual to find a place in the wild where a water course is dry in the wettest of years because of the way land above it has changed—for instance, when the watershed has been blocked by a rockfall or earthquake. In the geologically active west (we like to say it that way because it does not make our hearts beat fast the way it does when we say "on the earthquake-prone Pacific Rim") these changes happen quite suddenly. In 1992 a tremor knocked down houses and changed river courses in Humboldt County. And the wisdom of

Jon Stoumen on Secondary Performance

When we look at the way a building goes through its normal responses to its climate and regional environment as well as the way the residents live, their uses on a day-to-day basis, we are evaluating the house's primary performance.

We must never forget secondary performance, which is what we see when we look at the way buildings fall apart. Buildings we thought were built in really benign kinds of climates, if left long enough (especially unmaintained) finally fail, and by looking at how they fail, we can learn to make better buildings. An example: When there's a big fire, study the houses that survive, to see what their builders did right. A problem with modern architecture is that too many buildings of a certain genre get built without getting tested, like Maybeck's houses in Berkeley, with nice pretty shingles. The ones in the hills all burned up in the Berkeley fire in the 1930s. Another example is the devastation that hurricane Andrew did to minimal-code-standard buildings in Florida.

Secondary performance is how a building responds to severe environmental stress. A building has to perform in its at-rest state, and also withstand encounters with severe stress, much the way most organisms have to perform. If we look very carefully at successful regional buildings after one, two, three centuries, we can understand why some buildings are still standing when others have fallen. In many cases we cannot build them in exactly the same way as they were built before, so we knowingly substitute a modern response to the same stress. When we look at the old buildings in New Mexico, built out of thick adobe walls with interlocking vegas that run all the way through to the outside and beyond, we see a simple, appropriate response to primary and secondary challenges. We who work in California are concerned that adobe structures aren't good in earthquakes because we have seen some col-

lowland development in Florida, along the Gulf, and on Kauai's shorelines is being questioned by planners and insurance companies after devastating hurricanes in recent years.

In what we like to call modern times, humankind has presumed to manage the planet's surface and its energy flows. As self-appointed masters of the planet, we are resentful when a natural geologic process like a landslide disturbs our modes of transportation. Government willingly spends large sums to resist the forces of geologic change and clean up afterward when we build athwart a natural danger zone. Humans insist on their right to go anywhere and do anything at any time, and are willing to pay large fees and engage in earth-changing works to do so. People who live in independent homes want to transform that consciousness, and to live more in concurrence with nature's powers, because we have found that efforts to tame her are too expensive and unreliable.

The cataclysms of our planet's greatest and most unpredictable cycle, its great storms and droughts, fires and pestilence, and the periodic lapse. But some old adobe structures survive many earthquakes while others fail. The difference may be in the quality of construction—the wall-to-opening ratio, for example. The old churches and buildings that have survived have very few, small openings. More modern adobe structures may have giant openings, and thinner and taller walls relative to their thickness, and these are the ones that fail.

I got this idea outside a discount store in Marin. We had just been in Maine, where we had ridden kayaks on the ocean, and it was totally cool. O'Malley was inside getting some stuff, and there were two kayaks on the roof of the next car. It turned out the kayaks were owned by kayak instructors, so I asked them, what's a good kind of kayak to get? One instructor said, "The bewildering thing about buying kayaks is, if you buy a kayak that's stable in the kind of water that you learn on, benign conditions, and you want to take that kayak in rough conditions, it won't perform well. And if you buy a kayak that will perform well in rough conditions, then it won't perform well for beginners in easy conditions."

I started thinking about those two kinds of performance. A house doesn't always exist in rough conditions, and it doesn't always exist in mellow conditions, so the trick is to make one that will exist in both conditions and will exhibit characteristics that will be cool all the time. One of the great things about the old indigenous architecture is, they didn't have access to a great number of materials and techniques to build with. Take an eskimo kayak for example: you've got these people living on the top of world in super, unbelievably cold conditions, and they were able to build with few resources and a lot of knowledge totally sophisticated, streamlined, lightweight, stronger-than-shit boats, that they trusted their butts to in the most severe conditions in the world all the time. This is a boat-building tradition passed down from generation to generation. Kayaks are almost an architecture. They have to perform, they have to be warm, dry, flexible, light.

So the aim is to build a house that, like the kayak, has good performance characteristics. We want a house that can perform well in still water and also perform well in white water.

yawning and shuddering of the ground beneath us, are evidences that the earth is still inventing itself. There may be a periodicity to these catastrophes, but it is too long and intricate for us to comprehend. When we choose a home, we hope that our site is blessed with immunity from the upheavals of an unfinished planet, but we would be foolish not to give some thought to the disasters characteristic of our general region, in hopes of learning from older failures and successes, and of finding in traditional design and new materials the ways to insure that our houses, with a little help from us, will carry us through the chaos. Just as we try to understand the shifts in political weather, we must attempt to attune ourselves to the natural processes and befriend them, or, in the worst case, anticipate nature's tantrums and head for higher, sounder ground.

CHAPTER 14

SOLAR NEIGHBORHOODS

AT ECOTECH, AN INTERNATIONAL CONFERENCE ON ECOLOGY AND TECHNOLOGY that was held in Monterey, California, in November 1991, questions of neighborhood and community became unexpectedly heated topics of discussion. We could agree on little more than this: neighborhoods and communities, at least in the United States, are seriously at risk, and something should be done soon.

Like other institutions in the late twentieth century (marriage, parenting, and family, to name an important few), community life has been obscured by our need to survive, and neighborliness has been drowned out by television. We remind ourselves to "think globally, act locally" because we are overwhelmed by our insignificance in the face of regional, national, and global events, and seldom have time to act at all. Traditional communities and neighborhoods have fallen into disarray because they require time, intention, and effort on the part of their members. A collection of people must mean to make a community, and must work together to keep it, or entropy triumphs. Gated communities and neighborhood "crime-watch" programs fall far short of the friendliness we would like to find in small-town America.

At the same time, nontraditional communities, often incorporating global membership, are thriving. As you might expect, independent homesteaders, having stepped out of the conventional definitions of home life, are finding ways to gainfully re-explore the meanings and functions of communities, as well.

Cultural landscape: Arnie and Maria Valdez's story

Arnold Valdez and Maria Mondragon-Valdez are seventh generation residents of Colorado's San Luis Valley who live with their five children in a passive solar adobe home they designed and built. Their resource center, the Peoples Alternative Energy Service, is part of the compound where they live; from this center they share their findings about low-cost energy with their community.

Maria: Living in a solar house, you're like a cat—you track the sun. The kids follow the comfort, in the summer they spend more time downstairs, where it's cooler, and in winter, upstairs. You should have choices like that, not be trapped like in a house trailer.

I do my laundry according to the weather. If it's stormy, that's my break. And we take our baths at certain times . . .

Arnie: That's because our solar hot water is hottest in the late afternoons.

Our children are very important to us. We have five. One has finished high school and is working and going part-time to college, then we have three in school, and an infant nineteen months.

Maria: Only one was born in the hospital. The rest were born here: it's a way of bonding to the house. We like the sense of place we have here. People don't have that when they live in the standardized housing in a city. People move too much, and when they do, they lose their culture. You get attached to land when you're young, and if you're always moving, you lose those values.

Our families have been in the San Luis Valley since 1852, and were in Arroyo Hondo in the late 1700s, so we have a cultural landscape and agricultural customs that are deeply rooted.

Arnie: We try to instill our values and sense of place in our children, but you can't force it. A couple of our children won't leave, but some of them will. Hopefully they'll all return, even if only to retire. We went out for schooling, and saw a bit of the world, and then we came back. There's not much here to support advanced education, and so you have to create your own work, become self-employed, or fall back to doing what your people did before.

I work in the adobe tradition; this house was the first new adobe in ten years when we built it, and it reawakened a traditional style of building. Both of us had ancestors involved in building Fort Massachusetts, and then Fort Garland, the northernmost outposts of the federal war department in this part of the west. I'm coming around the circle, doing the same thing they did, but now I'm educating people and designing buildings: one evolution higher than my ancestors, but not different.

Maria: My great-great-grandfather—maybe I need to say one more great, I don't know for sure—was the first Hispanic to get a military contract for firewood. My dad has a propane business which my brother is taking over. So you see, my family has been in the energy business for a long time.

Our energy crisis started here in the late 1950s. Until then, we gathered firewood in the forests above the town, but the forest areas were closed to common use when a new owner purchased the land in the 1960s, and our traditional source was denied to us. Costilla County (that means rib) is considered the Appalachia of the west, one of the hundred poorest counties in the United States, and so the loss of that land for hunting, grazing, and woodgathering was very hard on our people.

Arnie: So now we turn to the sun for the energy we lost. We work with solar and environmental advocacy, working to develop regional self-reliance. And so our work continues to help us bond with the land.

Our resource center, the Peoples Alternative Energy Service, is a statement of energy self-sufficiency. From it, we do our community work, encouraging low-cost, energy-saving systems and advocating environmental issues. Because it's small and the overhead so low, and we're tenacious and committed, we've survived the Reagan-Bush years, during which so many solar organizations failed. Now we have momentum, and hope for a better future.

Maria: That was the building we lived in just after our marriage; then it was very cold, and we didn't know any better. When we built our house, we had better ideas about building solar, and it was reasonable to use power from the power lines for some appliances and light.

Arnie: Financing this house was a struggle, because it was adobe, and experimental. The Farmer's Home Administration stopped the funds once, and it was only because our senator intervened that we could continue. By doing so much of the work ourselves, keeping to the elemental concept that we were building shelter, and using found materials, we were able to keep the cost down to about ten dollars a square foot.

We originally had a 2-kilowatt wind generator and too few photovoltaic modules, and our system is connected to the grid so we can sell any extra electricity. The wind generator ran well for a few years, but the wind is so destructive, it sheered off the yaw column in one gust. Maintenance is tough because you have to climb the tower. We're in a marginal wind regime, and we got a wind machine meant for ten or more miles per hour; a unit meant for a slower regime might work better. We'd like to sell it, and enlarge our greenhouse instead.

In our community work, we teach from the resource center and our house, and use them as models and laboratory. We make mistakes first

Arnie and Maria Valdez doing spring planting in their in-home greenhouse.

here, on ourselves, and then we feel confident we can help people avoid similar pitfalls.

We plan to change the greenhouse to vertical glass, because it is more efficient, and will provide better space for us and the plants. We like the solar contribution and the diffuseness of the light, but the sloped glass must have the snow and dust swept off constantly.

State-of-the-art solutions are seldom suitable for low-cost applications, and so in the systems we design, the owner has to commit to maintenance. Because we work in the San Luis Valley with a heavy emphasis on small installations for low-income people in traditional communities, much of the best information comes from the Third World, where we work hard to keep up our network. The high-end information and applications are very interesting, but the best low-cost solutions are more common in Third World nations. It's difficult to connect, of course, because it takes money to get to the other communities and learn.

Maria: Because this is a poor county, outsiders think they can bring in their projects without asking and without opposition. Right now we are fighting a Texas-financed goldmine scheme which would spew cyanide and heavy metals into our water. We're citizen enforcers; we helped catch four noncompliant events, resulting in a permit violation fine and increased environmental monitoring. We fear the build-up of an industry like mining, because it would bring with it quick infrastructure, and spoil the local culture which has been building itself for so long.

The author and Arnie Valdez in front of the Valdez house (photo by Jon Stoumen).

Arnie: Of course, the community is split. They can all see that it might mean the end of the dream, but some see jobs, and think that's more important.

Maria: This issue, of holding the community together in the face of economic stress, is not so easy as solar.

Arnie: As soon as jobs are involved, the issues get cloudy.

Maria: This is the oldest community in Colorado. We're trying to keep the community intact, because it has a rare and endangered cultural landscape.

Arnie: Which needs clean water, clean air, and clean economic solutions to survive.

Maria: Who would want to build with adobe if it had poison in it?

Arnie: We believe that building security through a sustainable lifestyle is much surer than social security, which builds dependence. In San Luis, the Church is an important part of that life, and I'm looking for ways to work with the Church. We have tried to work with universities and government, but they send us specialists who are insensitive to the community's needs. That's why our "town square" is a checkerboard of paths to nowhere. HUD introduced energy-inefficient substandard housing in the 1970s: two-by-four frame constructions with quarter-inch plywood roofs, no insulation. Early in the 1980s, DOE put an active solar system on our museum, which is too complicated for us to fix, and so it's broken. The Church is much more a part of our community.

We have built an environmental center, and we hope to build a chapel on the hill above town, but this is a very fragile environment, and looking out over the whole San Luis area . . .

Maria: We're afraid we're shooting ourselves in the foot by creating such a spiritual center. Will people come, fall in love with the valley, and come back to buy, build, and gentrify our community? We don't want to market our cultural landscape. We want to maintain our cultural enclaves, because they are our treasures.

Architecture and buildings have an important role, because people respond to symbols. We are trying to come to grips with architecturally motivated tourism. There should be sacred space around cultural enclaves, and the cultures respected as national treasures; we can't tolerate resort development, subdivisions, mini-malls, and fast food chains.

Arnie: There's no building or land use planning code in Costilla County, and the only limit is the Farmer's Home Administration, which is the banker for low-income community homebuilders.

Maria: We'll be okay for a while, but regulation is coming. We hope for something enlightened, like Davis, California's alternative building code. As the oldest community in Colorado, we have to work to protect what we have, and the ambiance is working against us. We are inundated by people fleeing their cities. Aspen has already become a pseudoculture of second, third, fifth homes. In Santa Fe, inexpensive homes are now celotex boxes covered to look like adobe, and adobe is only for the wealthy. We are too proud to become the residential sacrifice area for the whole United States.

Arnie: "Capitalist Tool" Forbes's development, Trinchera Ranch, is a sample of this kind of profiteering: a huge site subdivided with no sensitivity to the land, eight hundred miles of roads leading nowhere, a jetport and conference center for the elite. When urban subdivisions are transplanted to a mountain setting, you cause plenty of trouble. Now erosion and wildlife problems make it clear what a bad idea it was.

Maria: Without a sense of vernacular architecture, for which purpose, and for whom the subdivision was made . . . It's a crime against nature and the community to develop like this. Will buyers, if they are lucky enough to buy a buildable lot, ask for help from people who know the land and weather?

Arnie: That's why our parish built the environmental center first even before the chapel, on the hill just above town: a simple, stand-alone structure where we'll offer an environmental view of the community. We hope it will merge ecology and spiritualism, and reinstill the environmental aspect of living with spiritualism. We hope visitors will get a sense of the environment, and maybe they'll leave with a sense of respect.

Bonding with the Land

The idea of working with our relationship to the land is not new. Our predecessors on the North American continent had an elaborate and caring relationship with every aspect, from rocks and shadows to the animals and each other. When a deer was taken, it was done because it was needed by the clan, and could be spared by the environment. Even then, it was taken as a loan, and a conscious effort was made to explain to the spirit of the land the reason for taking, and to beg forgiveness. This heightened sense of humility fell before the rifles and plows of pioneers and settlers who believed the land owed them a living. But the American land is powerful and still untamed, and human claims to dominion are shallow. As we mature, many of us remark upon the onset of an innate earth-oriented conservatism, which, when connected to the land, turns us into animists and worshippers of nature. Meanwhile, the greediest and most denatured continue to succeed in living entire lifetimes immune from these restorative influences.

The generation of children just now coming out of schools and commencing their own lives is the first generation of newcomers to the land to have lived from birth with an awareness that the planet has a finite bearing capacity, and a limited tolerance for greediness and gluttony; they are the first of a new race of American natives. These young ones struggle not to be angry at those of us who have gone before, at the raping and pillaging of their heritage, and the bequeathing to them of a darkening prospect of hard work and little luxury. Children born off-the-grid of parents who have taught themselves energy consciousness have an advantage in addition to their generally happy upbringings, in that conservation will be an ingrained habit. We are lucky, we older ones, that it is these young ones who will be caring for us and the planet in our dotage. It is important that we do all in our power to help them pull the rabbit out of the hat.

Land for the cost of a car: Nancy Hensley's story

Nancy Hensley lives in an off-the-grid home she built with her brother's help in the wooded hills of Greenfield Ranch, above Ukiah, California.

This story starts in about 1969 when my cousin was working on a dream he had of making it possible for good people to get good land in the country for the price of a new Volkswagen. His research turned up the Greenfield Ranch, a beautiful fifty-three hundred acre sheep and cattle ranch that wasn't making enough money to pay its taxes. With lots of help from friends who wanted to be a part of his dream, he was able to purchase the ranch. He divided it into smaller parcels, most of which have the essentials for country life: water, access, and a building site.

The Ranch has a history of being a thorn in the side of the county planning and building departments. About the time the county figured out we intended to subdivide the Ranch to the size parcels allowed by the ranch's original zoning, the county changed our zone. At that time, the county was already in trouble with the state, which was finding problems with the county's lack of a general plan. The state's pressure forced the county to impose a moratorium on all subdivisions from December of 1978 to September of 1981.

Building went on even though we couldn't subdivide. The county just didn't know how to deal with all the non-code homes that were popping up all over the Ranch. They couldn't just bulldoze them, even though they may have wanted to. After many years and long hours in the county supervisors' chambers, most of these problems were solved.

Nancy Hensley in the garden, practicing for retirement.

But now we are faced with the steeper fees of the 1990s, and the tougher standards for subdividing and building.

Back to 1971. My brother John, my parents, and I bought in as partners on two hundred acres. John and I moved onto the Ranch, set up a tent, built an outdoor dining table, pumped up the camp stove, filled kerosene lamps, hauled water from the spring, and started to build a house.

Getting outside power was out of the question. This was a pristine environment sparsely populated except for coyotes, raccoons, red-tailed hawks, magnificent madrones, and trees with names like Grandfather Fir. There was a power line running to the old Greenfield Ranch House at the center of the ranch, but when it was blown down in a storm we voted not to put it back up. (The Ranch has an active board of directors, and everything is done democratically.) The theme of our lives was simplicity and self-sufficiency. Years later, when my parents and I began plans for a new house, a hillside away, we looked again at the prospect of grid power. We could get outside power for about six thousand dollars. We all agreed that we could buy a lot of solar power for that much money, and were relieved not to get hooked into the same companies that create nuclear power.

John, along with my father, a neighbor, and two close friends, built our new home with passive solar in mind, but we didn't take the intensity of the summer sun into account, so we added a grape arbor to the south side of the house. It has made a big difference by shading the entire south side of the house in the summer and allowing full sun exposure in the winter. On cold winter days a good oak log keeps us warm for twenty-four hours. We burn less than a cord of wood a year unless we get an unusual amount of snow or icy weather.

Water is a big issue here, and fire is our biggest threat. The land divisions were made so almost every parcel has a water source. I pump my water from a 140-foot well up to two 1350-gallon tanks, then gravity-feed the house through more than half a mile of pipe. I use a gas generator and submersible pump, but I want to convert that to solar, soon.

When the Ranch does have a fire you really see what the word "community" can mean: everyone from miles around shows up to help.

Occasionally we have had problems with poachers hunting deer. We have the rare white deer here and we are very protective of them. The phone network kicks in as soon as there is a problem, and neighbors start setting up roadblocks. We wait for the fish and game officers to take care of "intruders." We're working on a solar entry gate, but it's a complicated problem because there are a hundred families living on the Ranch now.

We can use as much as a thousand gallons of water in one day when we are irrigating our large vegetable garden, boysenberries patch, flower garden, and fruit and nut tree orchard. When we aren't irrigating we can use as little as seventy-five gallons a day for household use. We use an inefficient propane batch water heater because we have too much calcium in our well water, which causes scale to build up too rapidly to allow us to use an instantaneous heater.

Our battery bank is 1600 amp-hours, retired from the telephone company. We have six 2-volt batteries weighing two hundred pounds each. They give us a good reserve of power in cloudy weather. The newest addition to our house is a 12-volt energy-efficient Sun Frost refrigerator that really needs two more solar panels added to our tracker to properly support its power needs. Without those extra panels we are finding ourselves low on power after three or four days of cloudy weather. Then we run the generator for a few hours to boost our battery bank back up.

We use 12-volt power upstairs so we aren't kicking on the inverter for lighting in the bedrooms. The downstairs is 110-volt house current for all our appliances in the kitchen and for entertainment in the living room.

It isn't obvious to guests that this house is independently powered, and they will often leave lights on. We need to be educate them about what it means to be your own power company. Living in a house like this takes getting used to. I like to be able to get a house sitter when we go away, but it can't be just anyone. We usually try to get someone from the community.

Even though I go to town almost every day to work, I find time for the garden, I like to think of it as my retirement plan, knowing how to grow my own food and chop my own wood.

Phoning It In: Telecommuting

When we take our skills to the end of the road, especially if we are trained as experts and specialists, we often join the ranks of the willfully unemployed. If we continue to commute to a distant job (merrily burning dead dinosaurs as we go) we have not yet fully committed ourselves to the whole proposition, and can hardly claim for ourselves an energy-sane lifestyle. The best recourse is to find a job that can be done at home, or within human-powered or mass-transit reach of home.

Creative jobs and cottage industries are satisfying sources of income where instead of commuters it is ideas, in the form of electrons or materials with a high value-to-mass ratio, that are transported back and forth. Many of the people who have told the stories in this book have figured out this employment problem, and others speak longingly of solutions such as working at home. Disconnecting is harder than it looks. Bosses who want to see bodies at desks, tasks that require personal interactions with co-workers or specialized equipment, the need for face-to-face service, and a whole litany of other reasons figure in the excuses we offer; all of these difficulties may in time be addressed, as the technologies of virtual presence and telecommunication become more familiar and accepted.

Many independent homesteaders are entrepreneurs. They have founded new forms of decentralized industry, self-reliant for energy and in every other way conscious of its possible environmental impact. These new ventures create alternative, dignified, and productive livelihoods for workers and artisans. Judging by my visits, this growing source of new employment must be considerable.

Making friends with energy: Michael Reynolds's story

Michael Reynolds is an architect working in Taos, New Mexico, originator of an innovative building technique using tires and tamped earth. The earthships which result are a modern application of the age-old southwestern adobe tradition. Because of their simple materials and great thermal mass, they are appropriate homes for their climate and for inexperienced owner-builders. And Mike is building more than houses: he is also building jobs and building community.

Michael Reynolds explaining his theory of energy.

It's simple observation: if you look up and see an avalanche coming at you, you get out of the way. That's common sense. For example, it's insane the way we build nuclear power plants and ship the electricity all over the country. Not only is there a threat from the plants, we have to deal with the crooked businessmen who sell and regulate the energy.

I came out of school interested in doing low-cost housing, and everywhere I turned, I ran into obstacles.

One night, probably more than twenty years ago, watching TV, I saw two specials, back to back: Walter Cronkite's piece on clear-cutting, in which he predicted a housing crisis in the near future (and he was obviously right about that), and Charles Kuralt predicting a waste crisis in which we would be buried by steel cans. It just fell into place for me: Take the source of the problem and turn it into a building material. It didn't take me long after that to patent my beer-can building block.

Once I've got an idea, my mind never stops. I started building houses out of cans. Then I saw piles of tires at the dump, and I thought, "energy crunch . . . thermal mass . . . " Nothing matches a three-foot rammed earth mass wall for energy efficiency. I could immediately see that, with simple skills and available, resilient materials, it would be easy to build cheap housing.

If you get into the right frame of mind to look at by-products, the answer is obvious. It's an inarguable natural phenomenon: we're running out of trees, but we've got big piles of used-up tires. If you were a spaceman, it's one of the first things you'd see. Here' s another obvious one: When we build a house, we're building a box to keep ourselves warm, right? Then we build a little box inside the bigger box, a refrigerator, and we bring in nuclear power to keep it cold. The spaceman would just shake his head. I'm working on a thermal mass refrigerator.

What you have to understand is, everything you need is out there. Now an earthship requires just a subtle amount of heat to stay comfortable through the winter. We've been putting in little three-hundred-dollar gas stoves that burn fossil fuel, but that doesn't make much sense. I started trying to think, where is heat? And how can I make friends with it, and get it? So we've come up with a little heater to provide just a little heat through the winter. It's a steel box; in September you put a bale of alfalfa in it, and wet it down and close it up. Pretty soon it's decaying, and it gives off heat through the winter, then in spring you open it up and turn the compost out into your garden. These little things can be built into benches, furniture, anything. Right now, we're working on finding out how many bales you need for how much heat.

I've been lucky in New Mexico. I hit the building department twenty years ago with the beer-can buildings, and I've been taking them along since, bending them little by little. The building code in New Mexico has changed with us, and helps us work with building departments in other states. You see, we figure that the code is an existing phenomenon, and we believe we can work with it.

For example, building officials don't like composting toilets, so we asked, why not? Well, they told us, people use them wrong, there's at least a dozen things you have to do to make them work right, and if you forget one, you end up with raw shit in the compost. That's dangerous. So we came up with the idea of our shit fryer. (We'll market it as the solar toilet, but that's not what we call it around here.) It's a special-purpose solar oven that cooks shit to powder at about 300 degrees in a couple of days. We took some of the powder and mixed it with drinking water, and sent it off to the health department, and the response came back: Lots of particulates, but sterile. We showed those results, and the fryer, to the inspector, and it made him happy.

Working with the inspectors, we hear their fear, and that becomes our design determinant. We solve it, and give them the solution; we simply disappear the problem they have. We get the solution to work, we don't care what it costs or how it looks at first, then we refine the solution, and build cottage industry and low-tech jobs for the people around here.

My whole life, I've seen inflation getting worse, and politicians chasing it across the sky. I aim to make it so that people don't need a job to survive. They need a job if they want a VCR or a Corvette, but if they want to reduce their needs to subsistence, they shouldn't need jobs. My overall objective isn't to create some architected environment; I'm trying a new approach, living past architecture into culture and economics. We talk about that in the Earthship books, how to get a house without the mortgage companies. We've made it happen here, up the mountain at Reach and out on the mesa at Star. It's just not right for people to go into hock for land, to pay big dollars for a teeny piece of land.

My idea is to get the profit out of land. For our first project, at Reach, we bought fifty-five acres of steep hillside, no power or water, and we let people buy in, a thousand square feet for a thousand dollars. They get free run of the rest of the land, and a support group of other like-minded people. Well, more people wanted in than there was space, so we started a much bigger project at Star.

As soon as we took infrastructure and profit out, took that weight off, the whole project rises to the surface like a cork, a buoyant success beyond our belief. It's not a subdivision, but more like a country club, which works okay in New Mexico because it doesn't trigger any of the zoning difficulties. Usually, a subdivision's big problems are infrastructure, sewage contamination, water table. We're not doing any of that; we're resting lightly on the land. All our water comes from roof catchments, so we're not using ground water. There are no septic tanks, and all our waste materials are recycled inside the shelter, so there's nothing that leaves the house. We're recycling grey water and not making any black water. This approach slips through the traditional boundaries and concerns. If we tried to build close in to a town, we might run into density restrictions, but not here in New Mexico or Montana. Our associations just haven't been challenged—it's like we're eighth-inch gravel passing through a quarter-inch screen.

The entrance requirement for one of these communities is environmental consciousness. Just wanting to build and live like this filters out the misfits. We get high rollers and low rollers, and they need each other. The high rollers are gone half the time, and need people to work on their houses and watch out, and the low rollers need the jobs. Maybe I oversimplify our response to the forces, but I think we've found a perfect way to work.

Let's say we want to grow flowers in the garden; there isn't an architect or scientist alive who can go out and put together the chemistry and mechanics of a living flower or a growing tree, but through our knowledge of science, we can make good soil. Well, the community is the plant, and it's made up of human nature and other things we can't make or change, and we know that certain kinds of soils grow certain kinds of plants. Like in a city, you can see why people are the way they are, because they're growing in spiky, dry, urban soil, and they need a lot of supervision, and are always looking for someone to take care of them. Now if we have luxurious soil, no stress because there's no debt for land and shelter, I trust the heart of the human beings. The community becomes the vessel that lets us transcend, and advance into our human

The nearly finished interior of Michael Reynolds's house at Star, near Taos, New Mexico.

potential. So you see, I'm not trying to design a community, I'm just building the right kind of soil . . .

In southern Humboldt County, sometimes lovingly called "the Solar Ghetto," in neighborhoods in New Mexico founded by Michael Reynolds, and elsewhere in the United States, inventive real-estate development schemes have liberated land from the grip of the grid, and enabled people of no great wealth to assume stewardship of land. Often, as on Frank Dolan's land, the profiteers have taken as much as they feel the land can give in their lifetimes, and are content to sell it to someone else. In other places, new developments are the result of population pressures or simply enlightened self-interest on the part of agents and regulators. Aided by the reluctance of power companies to expand their territories, as demonstrated by escalating costs and qualifications for line extensions, these new lands have often become solar neighborhoods. Those living in these largely unplanned communities often share little more than a willingness to produce or carry in their energy source. Like Wayne and Debbie Robertson's Wild West above Willits, this makes for some strange neighbors, but the spirit of the land often prevails, and the independence of self-reliant energy systems convinces even the most skeptical that home energy harvesting is possible and practical.

All remote neighborhoods share certain troublesome concerns. The access road is often the main problem, but government interference, water and timber rights, and differing opinions about the desirability of utilities, of wildlife and domestic animals, and of fire suppression, as well as a basket of other issues makes neighborhood life off-the-grid as complex as it ever was when neighborhoods were alive and well along Mainstreet, U.S.A. Many of us have forgotten how to be good neighbors, and government agencies are often literally beside themselves when it comes to dealing with nonconformity, yet the building of neighborhoods is an accelerating and increasingly dynamic aspect of life beyond the pavement and the power lines. There are lessons to be learned here by all of us who wish to restore neighborhood and family to the center of our lives.

CHAPTER 15

INDEPENDENT FUTURES

ALL LIFE ON THIS PLANET RELIES ON SOLAR ENERGY, FROM THE SIMPLEST BLUE-green algae right through to the whales and redwoods. Even rocks and waters use energy from the sun to make their changes. Humans are, to the best of our own knowledge, the only energy users with a hand and a voice in how energy gets utilized. In recent decades we have raised our voices and extended our hands to clamor and grasp for more and more energy. Our reach, which is global, and our greed, for which I have no words, have led us temporarily and spectacularly astray, costing other species their lives and rightful homes, and causing death, disease, and disarray among ourselves.

At the beginning of our nation, our ancestors huddling on the strands of a generous but forbidding land had no choice but to honor the spirits of the woods and mountains. The bravest of them wandered the woods, crossed the mountains, discovering great natural wealth wherever they went: the accumulation over millennia of the sun and its minions working to create magnificence. The people who lived in those woods and mountains, and their kin on the plains and deserts, were willing creatures of the sun, who walked in awe amidst wonders. The land endowed them with plenty, and by their habit and understanding, they easily lived within the sustainable limits of the land's wealth.

Growing in pride and population, the offspring of our early forebears, believing that this great land was theirs to take, began their westward progress, clearing the woods, slaughtering the game, and poisoning the streams, secure in the precept that the land was infinite. And if not infinite, it was immense, and so their depredations initially had only local effect; from the top of almost any hill, the prospect of virgin lands stretched to the horizon, promising that profit could be taken and errors could be left behind. The land was not always kindly, and so these settlers kept alive the pretense of man against nature, the heroic wresting of wealth out of wilderness. Who would stand against such heroism, especially when it provided riches and comfort to all who wished for it? Those who had tracked the land before, and who took a longer, subtler view of the world, retreated or were overwhelmed.

This pattern has continued. The Indians who had lived on the land for millennia were removed from the places that sustained them, and confined to ever poorer reservations, or assimilated, by force and by law. The land itself was vanquished, made to yield its timber, topsoil, and mineral wealth to the takers, and ravaged by the elements which degraded the wasted land with erosion and desertification and dust storm. The ancestral forest, the fragile species, even the mighty rivers were not proof against the onslaught, because the Euro-American, although puny in his singularity, had learned to hunt as an army, in numbers great enough to defeat nature. Having taken the easiest pickings on this continent, acquisitive developers turned to other continents, and other peoples, and worked their evil exploitation over and over again. A culture founded on quick seizure and rapid growth now finds itself against an elastic but impenetrable limitation: We have used up earth's easy resources and her ability to recover and restore.

Along the peripheries of our cultural reach, where the roads and power lines play out, and where we can still find a hilltop from which the vista seems limitless, the land retains its vigor, and nature's wisdom can still be sensed. We know, even as we resist such knowing with all the atavistic enthusiasm of the damned, that the limits are before us. Some who settle along the fringes maintain an eschatological view: "Humanity's headlong rush to devour nature has gone too far, and the end is near." To cheer us up, these prophets remind us that extinction is not rare on this planet, and the passing of our species, like that of the dinosaurs before us, is as ephemeral an occurrence as the wind in the trees, having no permanent importance in the larger scheme of things.

The voices in this book tell a different, more hopeful story. Explicitly, or between the lines, I hear them say that the tide is turning. Moving inward from the periphery now, we seek to pass back toward the center a renewed and regenerative partnership with the forces of Nature, a vision of the wisdom of the land, and a plan for salvaging the future for our children. We have found tools for the task: an approach to energy that cooperates with earthly processes rather than exploiting them, a conservative attitude toward matter and energy, and a collective sense of the value of all life. With this sense of partnership, this vision, and these tools we have, in the last few minutes of the eleventh hour, a chance to transcend and survive. The vista we cherish before us is the most exciting ever seen: the possibility that we may serve the people of the world, save the planet and all its life-forms, and explore the universe.

What We Know

We know that the way we use energy is critical to our success. If we continue to hurtle down the path of exhaustible energy, we will crash. If we can each slow our pace, develop an extraordinary regard for energy, and commit personally to living reasonably within the budget allotted to us by the sun, we may succeed.

We know that the tools are available to us. A solar-regulated life may be lived anywhere on this planet in comfort and plenty. We need not wait for technological miracles; the miracles we already have, if competently applied, will provide for all of us. Moreover, by learning to live with the energy supply offered freely by the sun, we elaborate a model than can, with few limitations, be refined and applied everywhere on the globe. We will find sufficient to our needs the sunlight that falls on us and on the plot of land we care for, and the winds and the waters that are moved by sun and seasons (as well as certain foreign and beloved necessities, including our chocolate and oranges).

The real limitations, though few, may be brutally inelastic. *Overpopulation*: We must learn that quality of life and the carrying capacity of an environment bear a constant relationship to each other; if we wish to improve individual life, we must learn to control the way we breed. Every other species on the planet does this, or has it done for them by natural controls. Having disabled the controls, we must learn to regulate ourselves or expect more powerful controls to assert themselves. *Greed*: We must learn that amassing wealth steals from others while spiritually impoverishing the wealthy. *Impatience*: We must learn to discern and wait for cycles, the fullness of time, and tread gently at first, being careful to seek and read the early signs that tell us if we are on course or gone astray. We must learn not to rush so quickly to market, but study the whole cost and effect of new endeavors. *Conformity*: We must learn to tolerate and encourage diversity in humankind and every other kind of life, for we never know when we will need to appeal to the hidden strengths that variation carries within its designs.

We know that the time available to us is not limitless. Accidents encountered along our present course have already frightened us into periods of terrorized immobility. Chernobyl, the Gulf War, AIDS, the ozone holes, global warming: signposts that we ourselves have erected, all bear the same small print: "Turn back; go no further; this way lies danger."

Quietly now, from the hills, down the muddy roads, the pioneers are coming back, reporting new-found discoveries and earthy delights. Not gold or timber, this time, nor peoples to enslave, nor fortunes to be made, but an even brighter hope: We have within ourselves, and falling from the sky, everything we need to live sustainably and well. This bounty from the energy frontier can be used to guide and enable a planetary renaissance.

What We Must Do

First, we must be rational in our own lives and in all our works. As we live and when we build, we must be attentive to the offerings of our environment, and live within the boundaries they suggest, right down to the smallest details. As we live with and within nature, and as there is no waste in nature, we must waste nothing. The voices in this book give good counsel. Wes Edwards recommends a small house. David Katz and Robert Sardinsky agree with Einstein's precept that "Everything should be as simple as possible, but no simpler." Felicia Cowden tells us that "Solar power's biggest gift to the environment is showing people that it is possible to live well without being wasteful." Hearing these words is not sufficient; we must put them to use.

Next, we must work to align ourselves, our neighbors, and our institutions with this course: the way of renewables. This will take all our patience, our intelligence, and our powers of persuasion. As opportunities for environmental exploitation and quick profit collapse, we must find ways to pay the whole cost of everything we buy—to make the marketplace responsive to long-term needs, as expressed by our children and grandchildren. But we must take an even longer view, beyond our familiar horizons, to the widening future we wish for our planet and all those living on it. By bringing the global vision home, Leigh Seddon, Steven Strong, Maria and Arnie Valdez, and, as you will soon read, Amory and Hunter Lovins urge us to participate, to help our leaders and representatives—who are more lost than we are—find the right way.

Finally, we must be lucky and brave. We can hope for serendipity, hope that we will find, by accident, the few tools and clues we need but do not yet have—a benign battery technology, a more efficient way to collect the sun's rays, a more compassionate and foresighted attitude toward the less powerful—but we cannot count on it. We must be vigilant, perpetually ready to take the better path when it's offered, and just as quick to backtrack when we see we have started down a wrong path. For example, we are sure now that the internal combustion engine, though familiar and addictive, is infernal; are we willing to give it up?

Rocky Mountain Institute

John Schaeffer, founder of Real Goods, leaves RMI after a visit with Hunter and Amory Lovins.

The Rocky Mountain Institute is a nonprofit resource policy group whose director of research, Amory Lovins, is an authority on energy strategy. The Institute's main offices are in a comfortable, super-insulated, four-thousand-square-foot rock, timber, and glass building beside Capitol Creek in Old Snowmass, Colorado. It gets cold at 7100 feet, down to −40 degrees every winter, but the Institute building has no conventional heating and derives 99 percent of its space heating from the sun shining into the bioshelter, a large greenhouse built into the building's center. The Institute's twenty-one desk office is on one side of the central bioshelter, while the kitchen, entryway, and Amory's residence and work pod are on the other. Energy-saving features are used throughout: the building's slipform, sixteen-inch-thick walls are R-40, and the roof is R-60; its windows are virtually all argon-filled double glazing with a thin-coated polyester film within, which lose less than a fifth as much heat as single-pane glass. The greenhouse roof and the windows in the curved east, south, and west walls allow winter light to reach the north wall to help keep the temperature uniform within the building. The glazed greenhouse roof continues above the roof line so that unwanted heat can be vented. Unusual and experimental features abound, including a solar clothes drying closet and air-to-air heat exchangers. From the copying machine to the kettles in the kitchen, every effort has been made to use energy efficiently.

The building feels like a busy, gracious house. The dining table is also the conference table, and visitors talk with staff there, between the kitchen and the greenhouse. The multiple open levels and curved walls give an extraordinary sense of the whole of the structure. On the north wall a recirculating waterfall, purportedly tuned by a master waterfall-tuner, feeds a small stream and two ponds in the greenhouse, which is filled with tropical plants and is home to an ill-tempered iguana; everyone else appears to be happy and productive.

The soft path: Hunter and Amory Lovins's story

From their mountain fastness beside Capitol Creek in Old Snowmass village, Hunter and Amory Lovins have been quietly working for many years to visualize and then realize a soft path for human survival through the hard realities of economic and nationalistic bullishness. By attending to their own home, and making the Rocky Mountain Institute a working model for the solutions they propose, they have become authoritative and experienced prophets. Their advice is sought by governments and corporations throughout the world. We talked, first about the house in which they and the Institute live, and then about the future.

Amory Lovins at the Rocky Mountain Institute (photo by Jeanette Darnauer).

Amory: In building the Institute, we tried to anticipate every threat. One nice catastrophe we're unprepared for would be an earth-sheltered elk on the flat roof falling through the greenhouse glass. We learned some important things in the uninsulated cardboard sieve of a house we were in before we moved here. A power failure, for example, caused a cascade of nasty events: with electricity out, we had no water, and the furnace stopped, and without water pressure or electric heat-tape, the pipes and radiant heating tubing soon froze.

In contrast, we tried to build our own place so there were three ways to do everything.

Hunter: It works well the way it is. We fix things that break, like the inverter, but we're not thinking of major changes.

Amory: There's no battery on the photovoltaic system, because we have a utility interconnect. Our electric co-op insisted on a grid-excited inverter, to guard against our energizing the line when they think it's not energized. Interestingly, our power has turned out to be of slightly better quality than theirs. If you regard this as a household, our 2-kilowatt array is two or three times what a very efficient household's requirement should be. The average use of the whole building is 1200 watts, but only a tenth of that is for the household; nearly all the rest is the office equipment in the Institute's headquarters at the other end of the building.

Hunter: This system is more hands-off than a normal house system, which takes quite a lot more ongoing maintenance.

Amory: But the maintenance required here is of a higher order, because the equipment is more intelligence-intensive.

Hunter: A house ought to have stand-alone integrity. This system is intelligence-intensive only because it was cobbled together—for example, the hot tub system. When it was built, the equipment was less common and more specialized, like house-trailer appliances, which, when they break, have to be sent out to be fixed. We now have local experts who come to work on our systems. For example, there are lots of people who can properly set the seasonal angle on the PV array or check the chemistry on the active solar water-heating loop.

One of my pet peeves is that so many people don't think about how passive systems work, and do stupid things. Even so, this house is pretty forgiving: if you leave a window open, the thermal mass reheats the overcooled air so the house temperature recovers.

Amory: The problems are often simple, like when the lower fishpond crashed, it turned out that algae had grown over the bubble hose. But we lost the fish for a few days until fry washed in the recirculating stream from the unaffected upper pond recolonized the lower one.

Hunter: And we have annual attacks of cotton scale . . .

Amory: Living here, we've learned that some of our original ideas can be further refined. We don't run the main air-to-air heat exchanger at night in midwinter, because letting the sun regulate the air is simpler and better. The greenhouse fan that heats up the CO_2-depleted boundary layer on the leaves (so the plants grow better) is usually connected to a solar module, and runs only when the sun warrants (although I notice it's disconnected now, which is one of the puzzles of having so many people in one house). We plan to run the waterfall (recirculating system)

L. Hunter Lovins (photo by Kathleen Menke).

the same way, with a 12-volt pump and dedicated PV, which would better match the pond's oxygen requirement. Things work quite well, even where right now it's partly manual and someone from the Institute has to occasionally turn a few things on and off.

Hunter: It is much busier here than it was ever meant to be. The building was planned for twelve people, but there are now thirty-six at the Institute, installed in six buildings.

Amory: We only have administration and outreach in this building now. Oh yes, and there's seventeen more at E source in Boulder . . . That's a lot of corks to keep underwater!

We're upgrading two of the other buildings that RMI owns, putting in high-tech glass. They all have efficient lighting, of course . . .

Hunter: We worry that governmental officials, who should know better, thoroughly confuse conservation and efficiency, and call for all Americans to sacrifice . . . Exactly the wrong thing to say; back to the days of Carter's sweater.

I wish I thought that *anybody* in the Administration really understood the fundamental problems outlined in, say, *Beyond the Limits*. I don't see anyone acting like they understand.

The new bunch may be a lot better than what has been in there before. It seems to me that most good people are going to say, "These new guys are so good, we don't have to do anything anymore." That is a risk that could get us all in a lot of trouble. There are enormous interest groups that want the business of government to carry on as it has for so long.

[*Hunter leaves.*]

Amory: I got a call today from a guy in the power generation business, who says that his industry has decided to lobby for the BTU tax to be levied *after* generation at the power plant, so that all the conversion losses go untaxed. Instead of taxing the fuel for 1000 BTUs, only 300 BTUs or so would end up getting taxed because the rest is upstream. Their precedent for this is that the refiners think they've got permission to pay on their output, not their input. The gas pipeline people want to pay at the retail, not the wholesale end. This is all breathtakingly dumb. The whole point is to tax depletable fuels as they come out of the ground or into the country, so you give an incentive to reduce not only end-use losses but also distribution and conversion losses.

I second Hunter's concern. Nixon made a wonderful speech about population, perhaps one of the best ever made by an American president, but nobody ever thanked him, and so he never mentioned it again. If a president tries to do something good, but gets opposition and not support, he'll back off. This also happened in the early Carter years, where we figured our people were in there now, we didn't need to work. We really need to work harder to make sure that the things we are trying to do actually happen. Efficiency and self-sufficiency issues now need public support, vigorously expressed.

I'm very interested in making changes around here that allow me to transmit electrons rather than protoplasm, using Picture Tel. You see, this protoplasm really doesn't like to travel. I've done Picture Tel for things like addressing a bunch of CEOs in Germany without having to go there, and it worked fine. There is a learning curve, and so the available equipment and software is decreasing rapidly in cost. The big price is still the equipment. We're having a terrible time with our telephone company, because we need upgraded service, and this is a rural system. I guess the local switches just can't handle it, so it may take a while.

About a seventh of all companies now accommodate telecommuting. At least seven million American workers already do their work at home, and their employers see a 15 to 20 percent gain in productivity when they do. Not to mention a lot of private and public costs in commuting avoided.

Electronic presence seems to be very well accepted as long as it is technically good. You need to be able to send graphics, and see what is going on. We are working on upgrading our systems, but it works very well to communicate with the Boulder office using fax and modem.

You're interested in the effects that a band of self-sufficient homesteaders might have on the global energy picture, yes? Small effects, I think, but material, and growing, and heading in the right direction.

It's realistic to plan on self-sufficiency if the desire is there, even if you include all the energies that go into a home. We certainly had it in mind when we built the Institute building. We used rock which is indigenous. The core of foam insulation and the Portland cement from Salt Lake made this an energy-intensive structure to build, but when amortized over such a long period, that cost is acceptable. The oak timbers from Arkansas were sustainably harvested, in that replacement oaks were planted.

I think a much broader question is, how will the utility system evolve?

Ultimately, it will look like the telephone company looked (when we had a telephone company): basically bookkeeping, dispatch, and operations mediating between a lot of distributed sources. Carl Weinberg, manager of research at PG&E, has an excellent concept of the distributed utility, in which you don't baseload the power plant, you baseload the grid. In the past, utilities would focus mainly on generation, less on transmission, still less on distribution, even less on customer service. In the future, if is going to be exactly the other way around. Generation has already become a commodity; PG&E just dissolved its engineering and construction division because they're not planning to build any more big plants. They are reshaping the company to match the new priorities of customer service and distribution emphasis. Even in more built-up areas where there is less of an economic or cultural incentive to go independent, you'll still end up with technically similar systems. When we ultimately do have cheap photovoltaics, they will be all over the place, and utilities will simply write off a lot of power plants and not build more.

Idaho Power and Light is planning skid-mounted PV systems of from 1 to 3 kilowatts to drop in wherever line extension would be uneconomic. Utilities in western and South Australia and even just outside the metro Sydney area are in a bind, because they are supposed to sell rural electricity within 10 percent of the urban price, even if it means extremely long line extensions. So they're starting to put in a lot of PVs. I think adding PVs to the mixture, along with end-use efficiency, makes a lot of sense.

The way we compare electric resources is all screwed up. There are about a dozen important advantages of dispersed, renewable sources (and usually of end-use efficiency, too), which are not being taken into account. The removal of fuel-price risk means that a lower discount rate should be used, which makes PVs cost-effective today. The lead-time difference by itself makes them worthwhile: you can add capacity with PVs much faster than by building a plant of any fuel type. And accounting for externalities makes PVs worthwhile, as does avoiding distribution.

An old friend of mine from Holland, Dr. Caes Daey Ouwens, built a 50 average-watt house, similar to ours except with a gas refrigerator. He then did a stand-alone PV system, instead of taking a drop off the pole a few meters away and installing a service entrance panel. So he was able to avoid a small part of the distribution cost. He couldn't avoid the cost of running wires down his street, but still he got a ten- to twelve-year payback from capital cost plus energy savings, which kind of surprised him. Then he went to a village in Indonesia that had sub-transmission running right by the village. There he was able to avoid the costs of drop, step-down, switch gear, transmission into the village, and wiring the village. Instead he gave everybody some efficient end-use devices and stand-alone PVs on amortization at the utility discount rate, without subsidy. The result: they got a positive cash flow immediately, because the debt service was less than the villagers were already paying for radio batteries and lamp kerosene.

If that's true for a Third World village, it ought to be true for a bunch of us, too. The Indonesian grid cost-per-kilowatt-hour is pretty low, and so the basic point is quite revealing. In December 1993, I'll be giving a paper on this whole subject, "Better Ways to Compare Electric Resources." There is a whole lot we are leaving out when we compare bus bar cost at a flat discount rate.

How long will it take for us to recognize this and put it into practice? Probably five to ten years. Some jurisdictions will probably do it earlier, which is why I want to write this paper this year, to put some pressure on.

I'm told that Real Goods has already helped twenty thousand homesteaders build off-the-grid homes. I think this will continue to happen in the private market.

The slowest-moving parts of the energy delivery apparatus are the regulatory agencies and the bulk of the utility industry. They are pretty much hand-in-hand at how fast they come along. The most progressive utilities, PG&E being the outstanding case, will be leading the rest. Florida Power is coming along; Tampa will take longer. There's been a surprising movement lately in Commonwealth Edison, American Electric Power, Southern Company. Some of the systems that used to be real stick-in-the-muds are now showing a lot of signs of intellectual life. We are seeing, I think, a generational change, as the people who put the old policies in place gradually die or retire. But some of the change is surely due to an incurable attack of market forces. It costs more, now, to back up and clean up grid power than it does to build a small, self-sufficient source. The FAA is using this approach to switch ground-control avionics over to PVs even when there is already grid power to the site.

Let me tell you some of the other things I see in the future. A photovoltaic panel with a micro-inverter built right into the frame, so you just connect it to the building's wiring. A variable selectivity window with photovoltaic sensors and drivers, also in the frame—distributed intelligence throughout a building, so you point a gizmo at it and push buttons to change the setpoints.

The biggest revolution will have to come in the building trades, because stick-building is so appallingly primitive. Good structural engineering dictates that a stud wall should be 15 percent wood, but generally it's more like 35 percent, because carpenters are so sloppy with corners and blocking.

The institutional barriers to such reforms are staggering. We've identified about twenty-five parties—in conception, finance, planning, building, maintenance, and so on *ad nauseum*—who speak different languages (there's no Rosetta Stone), and who have perfectly perverse incentives. The system penalizes efficiency and rewards inefficiency and obstruction, so we've got to find ways to turn that around, just as we did in the utility industry. The problem is, there's no customer feedback from end-users in housing design. We already know that lack of improvement is the death of any manufacturing business, and if we manufactured any product with as little customer feedback as when we make buildings, we'd have gone broke long ago.

We have the technical knowledge to yield buildings that are ten times as energy-efficient, and cost *less* to build, but we've seen hardly any real improvement in forty years. When an architect is studying, does he ever talk to a practicing mechanical engineer? He's too busy studying how to build something that is, as they say, all glass and no windows. When he's done, he pitches it over the wall to the MEs, and says, "Here, cool this!" Designing a building nowadays isn't team play, it's a relay race! We've got to find a way to make the rewards proportional to the performance efficiencies of the building. As it is now, fees and profits keep going up every time the baton is passed, and nobody's working or thinking too hard. That's obvious: if the building's equipment is simpler, there'll be less fee and less profit.

We think we've found a way to disconnect this loop: we're offering a two-part fee, first for design, then a bonus based on x percent of the energy savings. Ontario Hydro has a similar program where the design team gets a performance-based rebate from the utility equal to three years' worth of energy savings.

We've just signed off on the basic design for a new tract house in Davis, California, for PG&E, in which we were able to push the designers to see what it would take to get rid of the cooling hardware entirely. In the process of planning a modestly air-conditioned home, we challenged them to find, and set aside in a basket of cooling package measures, every technique, no matter how small, that might be used to reduce the cooling load further: double drywall, white roof with radiant barrier, superwindows, refinements on that order. When they'd finished the first design pass, they'd come up with a house better than Title 24 standards with one and a half tons of refrigeration instead of the three tons they expected. Then they shook out the cooling package basket, and found that it added up to just enough to make even the ton-and-a-half go away. But the best surprise was the overall cost: assuming that 36 percent of the initial electric use was user plug-loads that couldn't be analyzed until the occupants showed up—hair dryers, TVs, computers—they had realized a saving of 62 percent of the site energy, or 89 percent of the analyzed load, below the 1993 Title 24 standards. The marginal capital cost of the measures (compared with the base case) was minus two thousand dollars; and the marginal maintenance cost was minus seventeen hundred dollars. Win, win, win. Never mind that the house will be quieter, more comfortable, and more healthful . . .

Let me ask you something. How do you feel, meeting here in the Institute? Buildings are supposed to make you feel good, but too often the questions that get asked have to do with equipment, not with application. We made an effort to introduce curves into our interior space; if God meant us to live in boxes, wouldn't she have given us corners? The greenhouse's plants produce plenty of oxygen and ions and nice smells. And we tried to design out the mechanical noise and 60-Hertz electrical smog you find in almost every building. Hear that? That's pure silence (except for the waterfall, which is more or less tuned to alpha rhythm). If you knew you could live and work in space like this, would you be willing to work anywhere else?

To most of us, such things seem so obvious, and yet . . . my friends, we are in for difficult and dangerous times, but the consciousness is rising. We might yet get out of this alive . . . who knows?

The Future's Gifts

Throughout this book I have tried to keep an elemental, particle-sized focus on the work we are trying to do—making independent homes for ourselves. As each minuscule detail is perfected, and brought into a comfortable symbiosis with its neighboring details, an organic pattern of life grows up. As I travelled, my perspective widened, and I found these organic centers scattered everywhere I looked. Each is slowly, ever so slowly, extending its edges; where the edges touch, a strong, fibrous community begins to take hold, and a broader local awareness is born. This network is on the brink of consciousness—the hundredth monkey may not yet have joined hands, but she is standing in line.

In the next decade, we will see consciousness dawn as the sun rises. Possibilities which were unthinkable a quarter-century ago, and which become practical as I write and you read, will become the way it is done, everywhere. For two centuries, since the time when we learned to harness energy and use it to drive our tools, we have been assailed by restlessness and hunger. We found power so intoxicating that it has taken these two centuries to learn that power is not an imperative, but only a means. Still besieged by the lingering effects of our earlier greed, we must now come to rely on what resources remain to us; those who have already done so declare that the result is not impoverishment or hopelessness, but wealth and health.

Human values, natural values, will at last bring our focus home. There we will find ways to continue re-civilization of our society, to empower and enfranchise, and to broaden our regard for every form of life. We will put our restlessness and hunger aside, along with its trappings of national pride, race pride, sex pride, species pride, and be at last prepared to join the intergalactic family.

We need not look to heroic technology to bail us out; the time of the quick fix is finished, and we cannot count on miracles or *dei ex machinae* (gods climbing down from machines, the way Greek tragedies were resolved). Awakening from a two-century-long bad dream, where we made ourselves nearly insane with the belief that we could conquer nature with machines, we now know that technology grants tiny boons—a picture of the way a photon behaves, the splendor of *buckminsterfullerene.* We now have the necessary tools, and must find the ways to use them well.

The independent home is a masterpiece of appropriateness and hope. Over the next two decades, as the spirit of independence comes down from the mountains, pervades our culture, and is welcomed home by every family, we will all attain our energy independence.

A great weight will lift from the planet, and from our souls, wherein we feel our oneness with her, most powerfully—and we will be able to get on with the business of living.

Glossary

abate: a favorite word in the resource world, meaning to make something go away forever

AC: see alternating current

alternating current (AC): electricity that changes voltage periodically, typically sixty times a second (or fifty in Europe). This kind of electricity is easier to move

alternatively powered house: a house which gets its electricity or other power from unusual sources

ambient: the prevailing temperature, usually outdoors

americium: a radioactive element used in microscopic amounts in first-generation compact fluorescent bulbs and smoke detectors

amortize: calculation of short-term cost over a longer period

amp: measure of flowing electricity

amp-hour: measure of flowing electricity over time

architecture of dominion: building as if humankind ruled everything

audit: an energy audit seeks energy inefficiencies and prescribes improvements

avoided cost: the amount utilities must pay for independently produced power; in theory, this was to be the whole cost, including capital share to produce peak-demand power, but over the years supply-side weaseling redefined it to be something more like the cost of the fuel the utility avoided burning

Balance-of-system: equipment that controls the flow of electricity during generation and storage

baseload: the smallest amount of electricity required to keep utility customers operating at the time of lowest demand; a utility's minimum load

berm: earth mounded in an artificial hill

bioregion: an area, usually fairly large, with generally homogeneous flora and fauna

biosphere: the thin layer of water, soil, and air which supports all known life on earth

black water: what gets flushed down the toilet

boardfoot: standard measure of lumber: equivalent to a slab of wood one foot square and one inch thick

BOS: abbreviation for Balance-of-system

BTU: British Thermal Unit, the amount of heat required to raise the temperature of one pound of water one degree Fahrenheit; 3411 BTUs equals one kilowatt-hour

Bureau of Land Management (**BLM**)**:** a government agency and the largest landholder in the West, in recent years it has functioned with a little less integrity than the fox brings to mind the chickens

bus bar cost: the average cost of electricity delivered to the customer's distribution point

buy-back agreement or contract: an agreement between the utility and a customer that any excess electricity generated by the customer will be bought back for an agreed-upon amount

Capitalist Tool Forbes: the late Malcolm Forbes (aka Fiji Forbes), an unapologetic capitalist exploiter

carcinogenic: cancer causing

catalytic converter: a device attached to the exhaust of a vehicle or burner to help complete combustion

cat skinner: one who operates a bulldozer

catchments: tanks for catching and holding precipitation (rain or snow water)

Caterpillar: a company that makes bulldozers, or a character in *Alice in Wonderland*

celotex: a wood-based fiberboard

CF: compact fluorescent

CFCs: Chlorinated fluorocarbons, an industrial solvent and material widely used until implicated as a cause of ozone depletion in the atmosphere

clear cutting: a forestry practice, cutting all trees in a relatively large plot

code: as in "the code": The Uniform Building Code (which is not at all uniform) used as a blunt instrument to keep builders in line

compact fluorescent (CF): a modern form of lightbulb using an integral ballast

componentry: the collection of components selected and connected to perform a task, for example, to power an independent home

compost: the process by which organic materials break down, or the materials in the process of being broken down

conductance: a material's ability to allow electricity to flow through it; gold has very high conductance

controls: switches, valves, over-current protection (fuses and circuit breakers), and meters that enable us to manage energy systems.

conversion: changing energy from one form—for example, wind—to another—electricity; inverters and transformers change one "flavor" of electricity, like 110-volt house current, to another, perhaps 12-volt DC

core/coil-ballasted: the materials-rich device required to drive fluorescent lights; usually contains americium
cross-training: training workers to do each others' jobs
curb appeal: real estate terminology for how appealing a house is from the street
current: a predominant energy flow or flow of particles, as in a stream or an electric wire

days of autonomy: the length of time (in days) that a system's storage can supply normal requirements without replenishment from a source
DC: see direct current
dead dinosaurs: flippant term for fossil fuels
deciduous: opposite of evergreen, losing leaves in winter
decommissioning: power industry terminology for sweeping a radioactive power plant under the rug
degree-days: the sum, taken over an average year, of the lowest (for heating) or highest (for cooling) ambient daily temperatures
density restrictions: imposed by planners, the number of residences permitted on a plot of a given size. For example, my land is zoned for one single-family residence on a two-acre parcel
diesel: a fuel which powers simple internal combustion engines; it is cheaper than gasoline, and burns dirtier, and is therefore much beloved by the transportation and power industries
direct current (DC)**:** the complement of AC, or alternating current, presents one unvarying voltage to a load
distribution: the process and equipment associated with moving energy from where it is delivered to where it is needed; in simple terms, the wires and pipes in the walls
domestic energy production: the amount of energy produced by a household
door-snakes: long bean- or sand-filled socks that block the crack below the door
doping, dopant: small, minutely controlled amounts of specific chemicals introduced into the semiconductor matrix to control the density and probabilistic movement of free electrons
downhole: a piece of equipment, usually a pump, that is lowered down the hole (the well or shaft) to do its work
draft-inducer fan: a fan that forces a chimney to draw air.
drip irrigation: a technique which precisely delivers measured amounts of water through small tubes; an exceedingly efficient way to water plants

earthship: a rammed-earth structure based on tires filled with tamped earth; the term was coined by Michael Reynolds

efficiency: a narrow mathematical concept describing the proportion of a resource that can actually be converted into useful product or work; for example, sunlight falling on a PV module contains a given amount of energy, but the module can only convert a percentage of it into electricity

electric vehicle (EV): an automobile powered by an electric motor connected to batteries and/or photovoltaic panels; a nonpolluting replacement for dead-dinosaur driven cars

electrolyte: a liquid richly endowed with ions capable of reacting chemically with a battery's plates to store and release electricity

electromagnetic radiation (EMR): the invisible field around an electric device. Not much is known about the effects of EMR, but it makes many of us nervous

electronic ballasts: an improvement over core/coil ballasts, used to drive compact fluorescent lamps; contains no radioactivity

embodied: of energy, meaning literally the amount of energy required to produce an object in its present form; an inflated balloon's embodied energy includes the energy required to manufacture and blow it up

EMR: see electromagnetic radiation

endangered building materials: like sandalwood, koa, and teak; materials which, because of their beauty or resilience, have caused the endangerment of the species that produce them

energy: strictly construed, the potential to do work, that is, to move mass

energy independence: an entity or locale enjoys this when it is able to produce any energy it requires without active importation of a source

environmental assessment: a formal inventory of positive and negative pressures on the environment exerted by an entity

EV: see electric vehicle

excite the coils: a generator needs a permanent magnet or an electromagnet to induce a flow of electrons; in a grid-connected generator, energizing the primary electromagnet with power from the grid insures that, if the grid shuts down, the generator stops producing power

exotics: plant species imported from another bioregion

exploitive materials and techniques: like diamonds, luan, and teak; materials and techniques which rely on exploitation of humans or animals to maintain their competitiveness in the marketplace

externalities: considerations, often subtle or remote, which should be accounted for when evaluating a process or product, but usually are not; for example, externalities for a power plant may include downwind particulate fallout and acid rain, damage to life forms in the cooling water intake and effluent streams, and many other factors

extractive materials and techniques: nearly all fossil fuels and metals, as well as gypsum, marble, sand, gravel, and cement; materials and techniques which require mining, removal, and transportation from their source, which removal may cause dislocation and degradation at their source

feng shui: a Chinese term; Kaso is the Japanese term; both describe the geomantic study of house siting with respect to the directions and influences

ferroconcrete: a construction technique, an armature of iron contained in a concrete body, often a wall, slab, or tank

flatland: agricultural land, land without mountains

flatlanders: not a term of endearment; used by Vermonters and others to describe people that come from elsewhere

FNC: Fiber-Nickel-Cadmium, a new battery technology

foremothers: women who stood beside or behind the forefathers and told them what to do; results indicate that the forefathers didn't listen very well

fossil hydrocarbons: plant and animal material which, having been heated and compressed by overburden (soil, water, or rock) over a geologic period of time (millions of years) has turned to coal, oil, or gas

Frankenstein knife switches: large exposed power control devices, mostly antique

frequency: of a wave, the number of peaks in a period; for example, alternating current presents sixty peaks per second, so its frequency is sixty hertz (hertz is the standard unit for frequency when the period in question is one second)

gassifier: a heating device which burns so hotly that the fuel sublimes directly from its solid to its gaseous state, and burns very cleanly

generator: any device which produces electricity

genius loci: the genius of a place, its unique characteristics which distinguish it from all other places

gentrify: to increase the value of a community's land, usually artificially, by attracting and catering to a higher class of people than those already residing there

gotcha!: an unexpected outcome or effect. Example: carbon monoxide is an invisible and deadly ingredient in fossil-fueled automobile exhaust

gradient: a condition which changes with distance; for example, a trail up a hill has a gradient from its lower elevation to its higher. More often used for invisible conditions like temperature or electromagnetic force.

grandfathered: deemed legal by building and planning regulators by virtue of its existence before the building and planning regulations were put in effect

gravity-fed: a water delivery scheme which stores water with sufficient head (height) to produce adequate delivery

grey water: all other household effluents besides black water (toilet water); grey water may be reused with much less processing than black water

grid: a utility term for the network of transmission lines that distribute electricity from a variety of sources across a large area

grid-connected system: a house, office, or other electrical system that can draw its energy from the grid; although usually grid-power-consumers, grid-connected systems can provide power to the grid

grunt work: hard, unspecialized labor

haole: Hawaiian term for a non-Hawaiian

head: the distance water falls from intake to generator in a hydroelec tric system. Also the object on top of your neck.

headrace: in a hydroelectric system, the headrace leads water from the forebay to the penstock intake

heat exchanger: device that passes heat from one substance to another; in a solar hot water heater, for example, the heat exchanger takes heat harvested by a fluid circulating through the solar panel and transfers it to domestic hot water

high-tech glass: window constructions made of two sheets of glass, some times treated with a metallic deposition, sealed together hermetically, with the cavity filled by an inert gas and, often, a further plastic membrane; high-tech glass can have an R-value as high as 10

house current: in the United States, 117 volts root mean square of alternating current, plus or minus 7 volts; nominally 110-volt power: what comes out of most wall outlets

HUD: Federal Department of Housing and Urban Development

HVAC: Heating, Ventilation, and Air Conditioning; space conditioning

hydro turbine: a device which converts a stream of water into rotational energy

hydrometer: tool used to measure the specific gravity of a liquid

hydronic: contraction of hydro and electronic, usually applied to radiant in-floor heating systems and their sensors and pumps

hysteresis: the lag between cause and effect, between stimulus and response

incandescent bulb: a light source that produces light by heating a filament until it emits photons, which is quite an energy-intensive task

incident solar radiation (or insolation): the amount of sunlight falling on a place

indigenous plantings: gardening with plants native to the bioregion

Inductive transformer/rectifier: the little transformer device that powers many household appliances; an energy criminal that takes an unreasonably large amount of alternating electricity and converts it into a much smaller amount of current with different properties, for example, lower-voltage direct current

infiltration: air, at ambient temperature, blowing through cracks and holes in a house wall and spoiling the space conditioning

infrastructure: a buzz word for the underpinnings of civilization—roads, water mains, power and phone lines, fire suppression, ambulance, education, and governmental services are all infrastructure (*Infra*, is Latin for beneath); in a more technical sense, the repair infrastructure is local existence of repair personnel and parts for a given technology

insolation: a word coined from incident solar radiation, the amount of sunlight falling on a place

insulation: a material which keeps energy from crossing from one place to another: on electrical wire, it is the plastic or rubber that covers the conductor; in a building, insulation makes the walls, floor, and roof more resistant to the outside (ambient) temperature

Integrated Resource Planning: an effort by the utility industry to consider all resources and requirements in order to produce electricity as efficiently as possible

interface: the point where two different flows or energies interact; for example, a power system's interface with the human world is manifested as meters, which show system status, and controls, with which that status can be manipulated

internal combustion engines: gasoline engines, typically in automobiles, small stand-alone devices like chain saws and lawnmowers, and generators

inverter: the electrical device that changes direct current into alternating current

IRP: see Integrated Resource Planning

Jacobs windmill: a famous and largely antique device for turning wind into electricity

kaso: see feng shui

lanai: Hawaiian word for veranda or porch
latillas: Spanish term for small more-or-less straight branches which are placed over vegas to support a sod or adobe roof; the vegas are the backbones, and the latillas the ribs, that support the roof
leachates: the chemicals which leach out of solid materials when water falls on them; the leachates of a dump are not healthy
line extensions: what the power company does to bring their power lines to the consumer
line-tied system: an electrical system connected to the power lines, usually having domestic power-generating capacity and the ability to draw power from the grid or return power to the grid, depending on load and generator status
low pressure: usually of water, meaning that the head, or pressurization, is relatively small
low-emissivity (low-E): applied to high-tech windows, meaning that infrared or heat energy will not pass back out through the glass

malahini: Hawaiian term for newcomer
masonite: a compressed fiberboard bonded together with formaldehyde-based resin, manufactured by the company of the same name
ME: Mechanical Engineer; the engineers who usually work with heating and cooling, elevators, and the other mechanical devices in a large building
micro-climate: the climate in a small area, sometimes as small as a garden or the interior of a house; climate is distinct from weather in that it speaks for trends taken over a period of at least a year, while weather describes immediate conditions
micro-hydro: small hydro (falling water) generation
modules: the manufactured panels of photovoltaic cells; a module typically houses thirty-six cells in an aluminum frame covered with a glass or acrylic cover, organizes their wiring, and provides a junction box for connection between itself, other modules in the array, and the system
monocropping: planting a single species of plant over an unnaturally large area

net metering: a desirable form of buy-back agreement in which the line-tied house's electric meter turns in the utility's favor when grid power is being drawn, and in the system owner's favor when the house generation exceeds its needs and electricity is flowing into the grid; at the end of the payment period, when the meter is read, the system owner pays (or is paid by) the utility depending on the net metering

off-peak energy: electricity available during the baseload period, which is usually cheaper; utilities often must keep generators turning, and are eager to find users during these periods, and so sell off-peak energy for less
off-peak kilowatt: a kilowatt-hour of off-peak energy
off-the-grid: not connected to the power lines: energy self-sufficient
ohm: the basic unit of electrical resistance; I = RV, or, Current equals Resistance times Voltage
order of magnitude: multiplied or divided by ten; one hundred is an order of magnitude smaller than one thousand, and an order of magnitude larger than ten
overcurrent: too much current for the wiring; overcurrent protection, in the form of fuses and circuit breakers, guards against this

particulates: particles that are so small that the persist in suspension in air or water
passively heated: a shelter which has its space heated by the sun without using any other energy
patch cutting: clear-cutting (cutting all trees) on a much smaller scale, usually less than an acre
pathetic fallacy: attributing human motivations to inanimate objects or animals
peak demand: the largest amount of electricity demanded by a utility's customers; typically, peak demand happens in early afternoon on the hottest weekday of the year
peak kilowatt: a kilowatt hour of electricity take during peak demand, usually the most expensive electricity money can buy
peak watt: one-thousandth of the foregoing
pelton wheel: a special turbine, designed by someone named Pelton, for converting flowing water into rotational energy
periodic table of elements: a chart showing the chemical elements organized by the number of protons in their nuclei and the number of electrons in their outer, or valence, band
petro-chemical industry: the industry that extracts, refines, and distributes dead dinosaurs
petroleum: dead dinosaurs, or fossilized petrochemicals, in a convenient liquid form
phantom loads: "energy criminals" that are on even when you turn them off: instant-on TVs, microwaves with clocks; symptomatic of impatience and our sloppy preference for immediacy over efficiency
photon: the theoretical particle used to explain light

photophobic: fear of light (or preference for darkness), usually used of insects and animals; the opposite, phototropic, means light-seeking.
photovoltaics (PVs): modules which utilize the photovoltaic effect to generate useable amounts of electricity
picture-tel: a telephone connection, broader band than normal, capable of transmitting a televised image of a conversation's participants; a way to move information rather than protoplasm
piezo-electric ignition: a technology that produces an electric spark from a mechanical act, like turning a knob
piñon: a small form of pine common in the southwest; the source of pinenuts
pioneers: berserkers who leave the beaten path
plug-loads: all the little devices we plug into the walls
plutonium: a particularly nasty radioactive material used in nuclear generation of electricity; one atom is enough to kill you.
pollution: any dumping of toxic or unpleasant materials into air or water
polyurethane: a long-chain carbon molecule, a good basis for sealants, paints, and plastics
power: kinetic, or moving energy, actually performing work
power conditioning equipment: electrical devices that change electrical forms (an inverter is an example) or assure that the electricity is of the correct form and reliability for the equipment for which it provides; a surge protector is another example
PUC: Public Utilities Commission; many states call it something else, but this is the agency responsible for regulating utility rates and practices
PURPA: this 1978 legislation, the Public Utility Regulatory Policy Act, requires utilities to purchase power from anyone at the utility's avoided cost
PVs: photovoltaic modules

R-value: resistance value, used specifically of materials used for insulating structures. Fiberglass insulation three inches thick has an R-value of thirteen.
radioactive material: a substance which, left to itself, sheds tiny highly energetic pieces which put anyone nearby at great risk. Plutonium is one of these. Radioactive materials remain active indefinitely, but the time over which they are active is measured in terms of half life, the time it takes them to become half as active as they are now; plutonium's half life is a little over twenty-two thousand years.
ram pump: a water-pumping machine that uses a water-hammer effect (based on the inertia of flowing water) to lift water

rare building materials: building materials that are hard to find
rationalize: to make something rational (not make excuses). For example, to rationalize energy pricing, we must charge the whole cost including prospecting, extracting, transporting, refining, delivering, and abating the negative aspects of the foregoing activities
REA: Rural Electrification Administration, founded during the Depression
Real Goods: a Ukiah, California mail-order company providing a comprehensive selection of alternative energy equipment
rectify: converting alternating current to direct current; literally, getting the power on the "right" side of the zero line. (see the Wave Gallery)

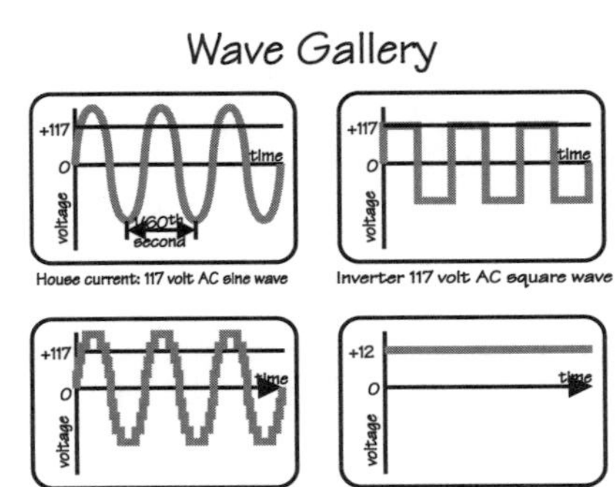

House current: 117 volt AC sine wave
Inverter 117 volt AC square wave
Inverter 117 volt AC synthesized sine wave
12 volt DC

regenerative capacity: the ability of the land to restore itself; two levels of regeneration can be discriminated, the first is when volunteer crops or planting mask the original damage, and the second is when the pre-existing flora re-establishes itself
release: a forestry term for opening up meadows that have been overgrown
renewable energy: an energy source that renews itself without effort; fossil fuels, once consumed, are gone forever, while solar energy is renewable in that the sun we harvest today has no effect on the sun we can harvest tomorrow
renewables: shorthand term for renewable energy or materials sources
resistance: the ability of a substance to resist electrical flow; in electricity, resistance is measured in ohms
retrofit: install new equipment into a structure which was not prepared for it; for example, we may retrofit a lamp with a compact fluorescent bulb
romex: an electrician's term for common two-conductor-with-ground wire, the kind used for houses
root mean square (RMS): the effective voltage of alternating current, usually about 70 percent (the square root of two over two) of the peak voltage; in house current typically has an RMS of 117 volts and a peak voltage of 167 volts

seduction by firelight: an enjoyable collegiate practice
Sequoia gigantea: the endangered larger, inland form of the California redwood
semiconductor: the chief ingredient in a photovoltaic cell, a normal insulating substance which conducts electricity under certain circumstances
SERI: Solar Energy Research Institute, a federally funded research establishment in Golden, Colorado
set-back thermostat: combines a clock and a thermostat so that a zone

(like a bedroom) may be kept comfortable only when in use.

siberian stove: a massive masonry stove; heated with an intense fire, the masonry retains and radiates heat long after the fire has gone out

sinker: a redwood log which has lain in water until waterlogged; for the builder, any saturated piece of lumber

sinusoidally: having a generally sine-wave shaped pattern; alternating current's voltage varies sinusoidally

site rose: the author's version of a feng shui diagram, showing the external factors which determine how a house should be sited

SMUD: Sacramento (California) Municipal Utilities District

solar aperture: the opening to the south of a site (in the northern hemisphere) across which the sun passes; trees, mountains, and buildings may narrow the aperture, which also changes with the season

solar fraction: the fraction of electricity which may be reasonably harvested from sun falling on a site; the solar fraction will be less in a foggy or cloudy site, or one with a narrower solar aperture, than in an open, sunny site

solar hot water heating: direct or indirect use of heat taken from the sun to heat domestic hot water

solar oven: simply a box with a glass front and, optionally, reflectors and reflector coated walls, which heats up in the sun sufficiently to cook food

solar panels: any kind of flat device placed in the sun to harvest solar energy

source: spring in French, in this book it is used to denote a place where energy originates or can be harvested

specific gravity: the relative density of a substance compared to water (for liquids and solids) or air (for gases): water is defined as 1.0; a fully charged sulphuric acid electrolyte might be as dense as 1.30, or 30 percent denser than water (specific gravity is measured with a hydrometer)

strategic building materials: materials which are useful in warfare and are not equally distributed about the globe, and therefore might be employed in nationalistic struggles. Petroleum is an example

strategic metals: metals useful in warfare and not equally distributed about the globe: copper, platinum, gold, and tungsten, are examples

sub-array: part of an array of PV modules wired together as a unit

subtransmission: secondary or feeder elements of the grid that distribute electricity from the major transmission lines to local communities

sustainable: material or energy sources which, if managed carefully, will provide at current levels indefinitely. A theoretical example: redwood would be sustainable if it is harvested sparingly (large takings and exportation to Japan not allowed) and if every tree taken were replaced with another redwood.

tailrace: in a hydroelectric system, the tailrace leads water from the turbine back to the stream
telephone batteries: telephones use DC voltage stored in large batteries; these batteries are well maintained and frequently replaced long before their design lives are over, and therefore often find their way into independent home installations
teratogenic: causing birth defects
therm: a quantity of natural gas, 100 cubic feet, roughly 100,000 BTUs of potential heat

thermal mass: solid, usually masonry volumes inside a structure which absorb heat, then radiate it slowly when the surrounding air falls below their temperature
thermography: photography of heat loss, usually with a special video camera sensitive to the far end of the infrared spectrum
thermopane: another term for high-tech glass
thermosiphon: a circulation system which takes advantage of the fact that warmer substances rise; by placing the solar collector of a solar hot water system below the tank, thermosiphoning takes care of circulating the hot water, and pumping is not required
time-of-day rates: electric rates that distinguish between electricity used during different times of the day; typically, more is charged for peak-demand times, and less for baseload times
transformers: a simple electrical device that changes the voltage of alternating current; most transformers are inductive, which means they set up a field around themselves, which is a costly thing to do
transparent energy system: a system that looks and acts like a conventional grid-connected home system, but is independent
troglodyte: cave dweller
trusses: a building technique which uses an engineered geometric construct of small members to replace a heavier timber
tungsten filament: the small coil in a lightbulb that glows hotly and brightly when electricity passes through it
turbine: a vaned wheel over which a rapidly moving liquid or gas is passed, causing the wheel to spin; a device for converting flow to rotational energy
twelve-volt (12-volt): a kind of direct current electricity, most commonly found in cars, but standard in independently powered homes

union: a plumbing part which allows disassembly without destruction by connecting two pipes together mechanically rather than with solder or glue

valence band: the outer, reactive cloud of electrons around an atom
VDTs: video display terminals, like televisions and computer screens
vegas: Spanish word for straight, round timbers, usually tree trunks, used to hold up a roof
video display terminals: televisions and computer screens
volt: measure of electrical potential. 110-volt house electricity has more potential to do work than an equal flow of 12-volt electricity

watt: measure of power (or work) equivalent to slightly less than one-thousandth of a horse power, just over three-thousandths of a BTU
wetlands: land which is marshy or wet in the rainy season (seasonal wetlands), or at high tide (tidal wetlands); until recently, wetlands were thought to be wastelands, and have been eagerly filled and converted to building sites, but now we find that a preponderance of lifeforms require wetlands in their life cycles (destruction of wet lands is an example of sinning in haste and never being able to repent, because wetlands are non-regenerative)
Whole-Life Cost Analysis: technique for evaluating all costs incident to a material or process, from its prospecting and extraction through manufacturing and transportation to its installation and use, and on to its final disposal. Often costs associated with prospecting, extraction, and disposal are left out of the price of a material or process, and are deferred to a distant place (as in acid rain) or time (our grandchildren's)
wind-spinners: fond name for wind machines, devices that turn wind into usable energy

xerophytes: drought-tolerant plants

Bibliography

Global issues:

Hall, Bob and Mary Lee Kerr. *1991-1992 Green Index.* Covelo, California: Island Press, 1991. State-by-state rankings of many environmental indicators.

Lovins, Amory B. *Soft Energy Paths.* Cambridge, England: Ballinger, 1977. Classic work on sustainable energy usage; the numbers are old (and paint a bleaker picture than our present, improved results), but the principles hold.

Meadows, Donella H., Dennis L. Meadows, and Jørgen Randers. *Beyond the Limits.* Post Mills, Vermont: Chelsea Green, 1992. An excellent analysis of resources and consumption, testing various courses using computer modeling, and concluding that, wi th effort and care, we can make it. A work full of heart as well as mind for the numerate reader.

Orr, David W. *Ecological Literacy.* Albany, New York: State University of New York Press, 1992. Essays on education for a sustainable society; excellent bibliography.

Rohter, Ira. *A Green Hawaii.* Honolulu, Hawaii: Na Kane O ka Malo Press, 1992. If the Green Party can win anywhere, it should be in Hawaii. This book explains what's wrong with island economics, then envisions the solutions.

Røstvick, Harald N. *The Sunshine Revolution.* Stavanger, Norway: SUN-LAB Forlag, 1992. A very pretty think book with many excellent quotes and a good vision of energy sanity from a northern European angle.

World Resources Institute. *World Resources 1992-93.* New York: Oxford University Press, 1992. A short but thorough review of global resources.

World Resources Institute. *The 1993 Information Please Environmental Almanac.* New York: Houghton Mifflin Company, 1993. Most of the facts you need to win an argument with an environmental boor.

Energy issues:

Perrin, Noel. *Solo.* New York: W. W. Norton, 1992. A thoughtful story about life with an electric vehicle.

Schaeffer, John, with friends and staff of Real Goods. *Alternative Energy Sourcebook.* Ukiah, California: Real Goods Trading Corporation, 1992. A catalog and compedium of alternative energy devices, tools and gadgets, and information about sustainable energy systems.

Wilson, Alex and John Morrill. *Consumer Guide to Energy Savings.* American Council for an Energy-Efficient Economy: Washington, D.C. and Berkeley, California: 1991. A small, authoritative, and comprehensive book about measures for energy efficiency. Excellent lists of energy equipment organized by relative efficiency. Technically correct and quite accessible.

Zuckermann, Wolfgang. *End of the Road.* Post Mills, Vermont: Chelsea Green Publishing Company, 1991. Why we need electric vehicles, and how we are going to get them.

Building:

AIA Research Corporation for The U.S. Department of Housing and Urban Development. *Regional Guidelines for Building Passive Energy Conserving Homes.* Washington: Superintendent of Documents, July 1980 HUD-PDR-355(2). Technical reference guide for regional climates and how to build using maximum energy efficiency.

Armstrong, Leslie. *The Little House.* New York: Collier Books, A Division of Macmillan Publishing Co., Inc., 1979. For those starting out on a limited budget, this book is organized to give you as much information as possible about design and construction and to encourage you to do as much of the work as you like. Drawings to scale.

Bodanis, David. *The Secret House.* New York: Simon & Schuster, 1986. This book will change the way you think about a house. Written from the house's viewpoint, it deals with tiny, particle-level events in the life of a house. Excellent tool for widening our perspective on what houses are and do.

Broome, Jon and Brian Richardson. *The Self-Build Book.* Bideford, Devon: Green Books, 1991. Very British, and therefore provides a good perspective on owner-builder practices in the United States.

Davidson, Joel. *The New Solar Electric Home.* Ann Arbor, Michigan: Aatec Publications, 1990. A thorough, technical treatment of photovoltaic self-sufficient home energy system design from a practical perspective.

Eccli, Eugene. *Low-Cost, Energy-Efficient Shelter.* Emmaus, Pennsylvania: Rodale Press, 1976. Low-cost housing designs and energy systems for the owner-builder.

Gault, Lila & Jeffrey Weiss. *Small Houses.* New York: Warner Books, Inc., 1980. The dreamers' pictorial of a variety of styles, intentions, and ideas of beautiful small houses.

Hylton, William H. *Build It Better Yourself.* Emmaus, Pennsylvania: Rodale Press, 1977. A complete homesteaders reference manual.

Kern, Ken. *The Owner Built Home.* New York: Charles Scribner's Sons, 1975. Text and sketches geared toward the alternative owner-builder. Detailed in representing the thought process of designing and building a home.

Marinelli, Janet. *The Naturally Elegant Home.* Boston, Massachusetts: Little, Brown and Company, 1992. A very pretty picture book of some spectacular natural homes.

McClintock, Mike. *Alternative Housebuilding.* New York: Sterling Publishing Co., Inc., 1989.

Meinert, David L. *Energy Conservation in Housing.* New York: Vantage Press, (need date).

Merrilees, Doug and Evelyn Loveday. *Low-Cost Pole Building Construction.* Pownal, Vermont: Storey Communications, Inc., 1990.

Pearson, David. *The Natural House Book.* New York: Simon & Schuster Inc., 1989.

Reader's Digest. *Complete Do-it-Yourself Manual.* Philippine: Reader's Digest Association for East Ltd., 1977. Excellent manual to all around home construction and repair.

Sardinsky, Robert, and the staff of the Rocky Mountain Institute. *The Efficient House Sourcebook.* Snowmass, Colorado: Rocky Mountain Institute, 1992. Inclusive listing of sources of information about energy efficiency organized by subject. Books, periodicals, and national agencies are reviewed, and books are often briefly excerpted. A very complete resource.

Schomer, Victoria. *Interior Concerns Resource Guide.* Mill Valley, California: Interior Concerns Publications, 1991. An excellent list of hypoallergenic and energy appropriate materials and sources.

Schwenke, Karl and Sue. *Build Your Own Stone House.* Pownal, Vermont: Storey Communications, Inc., 1991.

Strong, Steven J., with William G. Scheller. *The Solar Electric House.* Still River, Massachusetts: Sustainability Press, 1987, 1993. Another thorough treatment of photovoltaic and energy efficient house design from a more mechanical and theoretical perspective.

Wagner, Willis H. *Modern Carpentry.* South Holland, Illinois: The Goodheart-Willcox Co., Inc., 1979. Easy-to-understand encyclopedia of information on building materials and construction methods.

Systems:

Architectural Energy Corporation. *Maintenance and Operation of Stand-Alone Photovoltaic Systems.* Albuquerque, New Mexico: Sandia National Laboratories, 1991. Very technical, encyclopedic guide.

Campbell, Stu. *The Home Water Supply.* Pownal, Vermont: Storey Communication, Inc., 1983. Thorough, and well-illustrated.

Cary, Jere. *Building Your Own Kitchen Cabinets.* Newtown, Connecticut: The Taunton Press, Inc., 1983. A tool for beginners on layout, materials, construction, and installation of kitchen cabinets. A good resource for craftsmen at all levels.

Etherington, Larry R. *Auditing Your Home Utility Bills.* Phoenix, Arizona: UAS Publications, 1991, and *How to Save Money on your Electric, Gas, Water & Telephone Bills!* Phoenix, Arizonia: UAS Publications, 1991. Two books that give you the insight and authority you need to audit your utility bills.

Glover, Thomas J. *Pocket Ref.* Littleton, Colorado: Sequoia Publishing, Inc., 1992. An indispensible little book full of building and electrical knowledge.

Matson, Tim. *Earth Ponds.* Woodstock, Vermont: Countryman Press, 1991. Excellent work on building natural ponds.

Ogden, Joan M., and Robert Williams. *Solar Hydrogen: Moving Beyond Fossil Fuels.* Washington, D.C.: World Resources Institute, 1989. An overview of emerging hydrogen technologies.

Rauch, Paul H. *How to be Your Own Contractor.* Acton, Massachusetts: Brick House Publishing Company, 1988.

Scher, Les and Carol. *Finding and Buying Your Place in the Country.* Dearborn Financial Publishing, Inc., 1992. Everything the inexperienced land buyer needs to know before heading out to the country.

Sunset. *Solar Heating and Cooling.* Menlo Park, California: Lane Publishing Co., 1978. The basics of active and passive systems, hot water heaters, pools, spas and tubs. Sketches and color photographs.

Vivian, John. *Building Stone Walls.* Pownal, Vermont: Garden Way Publishing, 1978.

Water Supply Subcommittee of the Midwest Plan Service. *Private Water Systems Handbook.* Ames, Iowa: Midwest Plan Services, 1979. An excellent guide for planning and installing private water systems.

Gardening:

Belanger, Jerome D. *Raising Small Livestock.* Emmaus, Pennsylvania: Rodale Press, 1974. Homesteader's handbook to raising, breeding, butchering and cooking small livestock.

Berry, Wendell. *The Unsettling of America: Culture and Agriculture.* New York: Avon Books, 1977. A good explanation of what went wrong.

Coleman, Eliot. *The New Organic Grower.* Chelsea, Vermont: Chelsea Green Publishing Company, 1989, and *The New Organic Grower's Four-Season Harvest.* Post Mills, Vermont: Chelsea Green Publishing Company, 1992. Wonderful books on year-around organic gardening.

Organic Gardening magazine staff. *The Encyclopedia of Organic Gardening.* Emmaus, Pennsylvania: Rodale Press, 1978. This is the bible for organic gardeners.

Resource List

Committee on the Environment
American Institute of Architects
1735 New York Avenue NW
Washington, DC 20006
(202) 626-7300

American Solar Energy Society
2400 Central Avenue, Unit B-1
Boulder, CO 80301
(303)-443-3130

American Wind Energy Association
777 N. Capitol Street NE, Suite 805
Washington, DC 20002
(202) 408-8988

Conservation & Renewable Energy
Inquiry and Referral Service
P.O. Box 3048
Merrifield, VA 22116
(800) 523-2929

Energy Conserving Passive Solar Houses
Drawing Room Graphic Services
Box 88627
North Vancouver, BC
Canada V7L 4L2
(604)-689-1841

Energy Efficient Builders Association
1000 West Campus Drive
Wausau, WI 54401
(715)-675-6331

Florida Solar Energy Research Center
300 State Road 401
Cape Canaveral, FL 32920-4099
(407)-783-0300

Independent Home News
Caspar Institute
14992 Caspar Road #88
Caspar, CA 95420
fax: (707) 964-8978

National Appropriate Technology Assistance Service
P.O. Box 2525
Butte, MT 59702
(800) 428-2525

Northeast Sustainable Energy Association (NESEA)
23 Ames Street
Greenfield, MA 01301
(413) 774-6051

Photovoltaic Design Assistance Center
Sandia National Laboratories
Albuquerque, NM 87185-5800

Public Citizen
P.O. Box 19404
215 Pennsylvania Avenue SE
Washington, DC 20036
(202) 546-4996

Real Goods Trading Corporation
966 Mazzoni Street
Ukiah, CA 95482
(800) 762-7325
(707) 468-9292

The Rocky Mountain Institute
1739 Snowmass Creek Road
Snowmass, CO 81654-9199
(303) 927-3851

Southwest Technology Development Institute
New Mexico State University
Box 30001, Dept. 3SOL
Las Cruces, NM 88003-0001

Solar Design Associates, Inc.
P.O. Box 242
Harvard, MA 01451
(508) 456-6855

Solar Energy International
Box 1115
Carbondale, CO 81623-1115
(303) 963-0715

Union of Concerned Scientists
26 Church Street
Cambridge, MA 02238
(617) 547-5552

U.S. Capitol building
switchboard: (202) 224-3121

United States Department of Energy
100 Independence Avenue SW
Washington, DC 20585
(202) 586-6210

United States Environmental Protection Agency
401 M Street SW
Washington, DC 20460
(202) 382-4700

Worldwatch Institute
1776 Massachusetts Avenue NW
Washington, DC 20036
(202) 452-1999

Your Senator
Senate Office Building
Washington, DC 20510

Your Representative
House Office Building
Washington, DC 20515

A more complete listing of resources on all aspects of renewable energy can be found in the *Real Goods Alternative Energy Sourcebook, 7th Edition* (see the listing for Real Goods, above).

Index